SpringerBriefs in Systems Biology

For further volumes:
http://www.springer.com/series/10426

Ali Masoudi-Nejad · Zahra Narimani
Nazanin Hosseinkhan

Next Generation Sequencing and Sequence Assembly

Methodologies and Algorithms

 Springer

Ali Masoudi-Nejad
Laboratory of Systems Biology
and Bioinformatics (LBB)
Institute of Biochemistry and Biophysics
University of Tehran
Tehran
Iran

Nazanin Hosseinkhan
Laboratory of Systems Biology
and Bioinformatics (LBB)
Institute of Biochemistry and Biophysics
University of Tehran
Tehran
Iran

Zahra Narimani
Laboratory of Systems Biology
and Bioinformatics (LBB)
Institute of Biochemistry and Biophysics
University of Tehran
Tehran
Iran

ISSN 2193-4746   ISSN 2193-4754 (electronic)
ISBN 978-1-4614-7725-9 ISBN 978-1-4614-7726-6 (eBook)
DOI 10.1007/978-1-4614-7726-6
Springer New York Heidelberg Dordrecht London

Library of Congress Control Number: 2013938267

Printed on acid-free paper

Springer is part of Springer Science+Business Media (www.springer.com)

Dedicated to our loving family

Preface

DNA sequencing is a fast-moving science with technologies and platforms being updated at breathtaking speed. The hallmark of next generation sequencing (NGS) has been a massive increase in throughput and a decrease in price compared with previous technologies. The first next-generation DNA sequencing machine was introduced to the market by 454 Life Sciences (Basel, Switzerland) in 2005. The technology is based on a large-scale parallel pyrosequencing system, which relies on fixing nebulized and adapter-ligated DNA fragments to small DNA-capture beads in a water-in-oil emulsion. The Illumina's (CA, USA) Genome Analyzer was released in 2007 and marked a true revolution for genome sequencing in which short reads became significant to genomic applications. The technology is based on reversible dye terminators. DNA molecules are first attached to primers on a slide and amplified so that local clonal colonies are formed. Life Technologies' (CA, USA) SOLiD^{TM} technology employs sequencing by ligation. In this technology, a pool of all possible oligonucleotides of a fixed length is labeled according to the sequenced position. Oligonucleotides are annealed and ligated; the preferential ligation by DNA ligase for matching sequences results in a signal that is informative of the nucleotide at that position.

So-called 'third-generation' technologies directly sequence individual DNA molecules rather than relying on any amplification prior to sequencing. The recently released PacBio system can produce 35–45 Mb of data per cell with an average read length of 1,500 bp. The Ion Torrent Personal Genome Machine (PGM) is another third-generation platform that uses standard sequencing chemistry, but with a novel, semiconductor-based detection system. This technology already claims read lengths of approximately 200 bp with high accuracy, and the latest PGM 318 chip can produce 1.0 Gb of data in a 2-h run. When the implications of NGS technology became apparent, several assemblers were designed to deal with the new problems, i.e., assembly of short NGS reads in order to reconstruct the main longer sequences. Assembly process can be done either having a reference genome available (mapping) or without having a reference genome available (de Novo assembly). De Novo assembly algorithms, discussed in more detail in this book, can be classified into three main categories: greedy algorithms, Overlap-Layout-Consensus (OLC) methods, and De Bruijn graph approaches. The Euler assembler was the first to employ de Bruijn graphs for whole genome shotgun (WGS) assembly, and proved capable of assembling bacterial genomes. Velvet and ALLPATHS improved assembly in terms of speed, contig and scaffold length, and avoidance of misassembly. ABySS followed the innovations with de Bruijn methods, but also introduced a distributed representation of the graph, allowing message passing interface parallelization. The CABOGand variant MSR-CA pipelines are updates of the Celera overlap-based assembler designed for a combination of read types, which showed some success with short-read data for genomes in the 100 Mb range. The String Graph Assembler (SGA) is the first to make assembly of mammalian-sized genomes practical using the string graph approach. This observation on the current tradeoff between accuracy and continuity suggests avenues for future improvements in assembly. There is room for other improvements at the scaffolding stage, where, as has happened at the assembly stage, we witness a move from naïve and greedy algorithms to more subtle graph-based techniques.

In this book, we briefly introduce the history of first, second, and third generation sequencing technologies and also describe drawbacks of the old techniques which now are not suitable due to their cost and the need for automation which could not be achieved in those methods. In Sect. 2 major NGS methods—namely Roche/454 FLX, Illumina/Solexa Genome Analyzer, and Applied Biosystems SOLiD System, etc.—are described in detail. Also, after bringing the latest and most predominant technologies in NGS, nanopore DNA sequencing and Pacific single molecule real time (SMRT) DNA sequencing, which does not need an amplification step, are described. Latest subsections of this section are devoted to information about sequencing costs, file formats of the output, a comparison of methods, and their drawbacks, and finally application of NGS technologies. The second two sections, i.e. Sects. 3 and 4, provide an overview of the algorithmic view of the assembly problem. Our main focus in these two sections will be on de Novo assembly algorithms of NGS reads. In Sect. 3, we generally define the assembly problem and mention the challenges involved in the assembly process, including errors propagated from sequencing process beside computational challenges. Appropriate use of paired-end read data, which helps to overcome the challenges regarding short length of reads, and also preprocessing that helps to eliminate some other issues regarding inaccurate data, is the next topic discussed in this section. Using all these techniques to reduce problems, there will still be errors in assembly, and relevant assembly algorithms are needed to be validated in a standard way: These are the final topics which will be discussed in Sect. 3. Finally, in Sect. 4, an exact view of the assembly algorithm is given as to how the problem can be mapped to a graph and how different kind of graphs are treated in finding the solution, which is the final assembled genome. Concerning each of the assembly approaches, several example algorithms are then described in detail and, finally, a comparison of these methods is provided in Sect. 4.

Contents

Chapter 1
Next-Generation Sequencing Methodologies

1.1 Introduction

1.1.1 A Brief History of the Discovery of DNA Structure and Function

Although many people believe that the American biologist James Watson and English physicist Francis Crick were the first to discover DNA in the 1950s, DNA was actually discovered by the Swiss chemist Friedrich Miescher in the late 1860s during his attempts to isolate the protein components of leukocytes. But when he isolated a substance that was unlike proteins resistant to proteolysis and also had different chemical properties of proteins, including a much higher phosphorous content, he realized that he had discovered a new substance [1]. He called this new substance "nuclein."

Miescher's finding was not considered particularly important until the twentieth century, when the chemical nature of nuclein was studied by the Russian biochemist Phoebus Levene. He was the first to discover: (1) the order of three major components of a single nucleotide (phosphate-sugar-base) (Fig. 1.1); (2) the carbohydrate component of RNA (ribose) and DNA (deoxyribose); and (3) the way RNA and DNA molecules are put together. In 1919 Levene proposed that nucleic acids were composed of a series of nucleotides and that each nucleotide was in turn composed of just one of four nitrogen-containing bases—a sugar molecule and a phosphate group.

Studies conducted to discover the DNA structure were continued by Erwin Chargaff, an Austrian biochemist, to uncover additional details about the structure of DNA. He reached two major conclusions [3]: First, he stated that the nucleotide composition of DNA varies among species, and second, he concluded that the amount of the base adenine (A) is usually similar to the amount of thymine (T); this is also true about the amount of guanine (G) and cytosine (C). The latter is known as Chargaff's rule (Fig. 1.2).

A. Masoudi-Nejad et al., *Next Generation Sequencing and Sequence Assembly*, SpringerBriefs in Systems Biology, DOI: 10.1007/978-1-4614-7726-6_1, © The Author(s) 2013

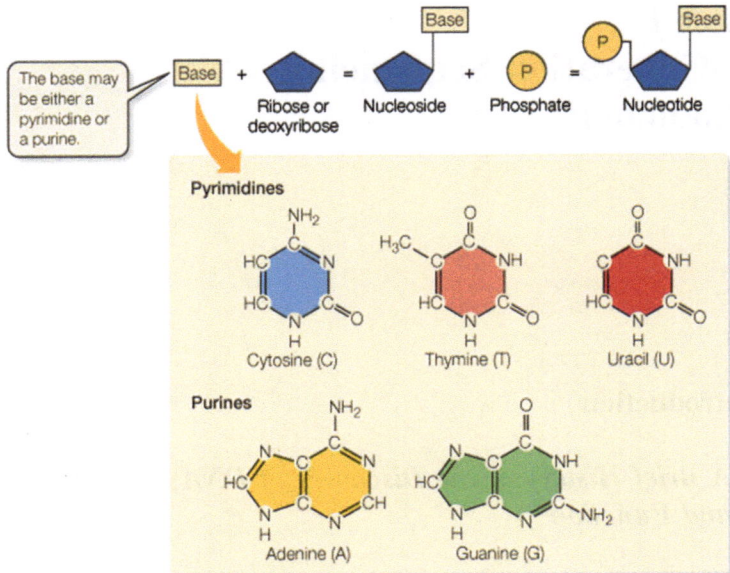

Fig. 1.1 Three components of each nucleotide: the nitrogenous base that can basically belong to two categories (*single ring*: pyrimidines, or *two-linked rings*: purines), a pentose sugar (ribose in RNA and deoxyribose in DNA), and a phosphate group [2]

Fig. 1.2 Chargaff's rule: the total amount of purines is equal to the total amount of pyrimidines [2]

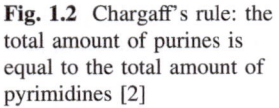
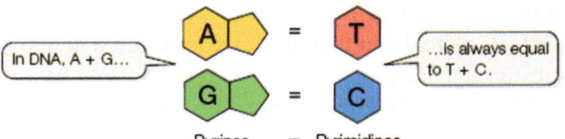

Chargaff's finding that A = T and C = G, along with some vital crystallography results obtained by the English researchers Rosalind Franklin and Maurice Wilkins, established a strong basis for the discovery of a three-dimensional, double-helical model for the structure of DNA proposed by Watson and Crick (Fig. 1.3).

Each chain of a double-helix DNA molecule is made up of the phosphodiester links between nucleotides. Two strands of a DNA molecule have different directionality. The two different ends of a single strand are called 3′ and 5′ and the direction of DNA synthesis is 5′>3′; this means that the free 3′ hydroxyl (OH) group from the growing strand of DNA attacks the phosphate on the next base to be added (Fig. 1.4). Pyrophosphate is released and the new base forms a phosphodiester bond with the growing strand of DNA. The free 3′ hydroxyl group is then free to attack the next base to be added. This reaction is catalyzed by DNA polymerases.

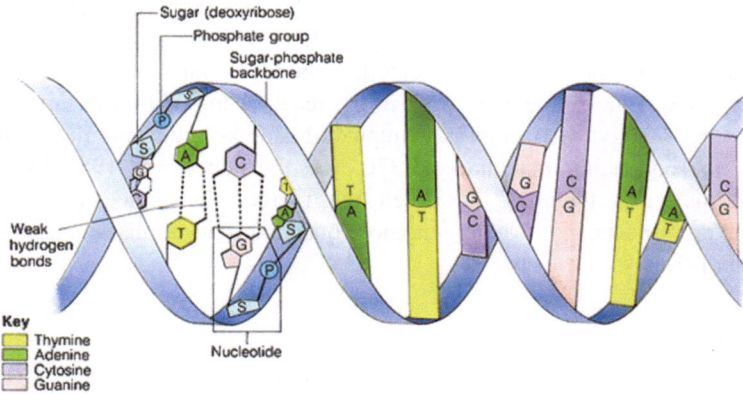

Fig. 1.3 Double-helical structure of DNA. The chains of sugar-phosphate groups are linked together by complementary bases [2]

Fig. 1.4 DNA synthesis direction. The 5′ end of the new nucleotide is linked to the 3′-OH of the last nucleotide of the growing chain by DNA polymerase action. During this reaction, a pyrophosphate group is released [http://www.prism.gatech.edu/~gh19/b1510/dnarep.htm]

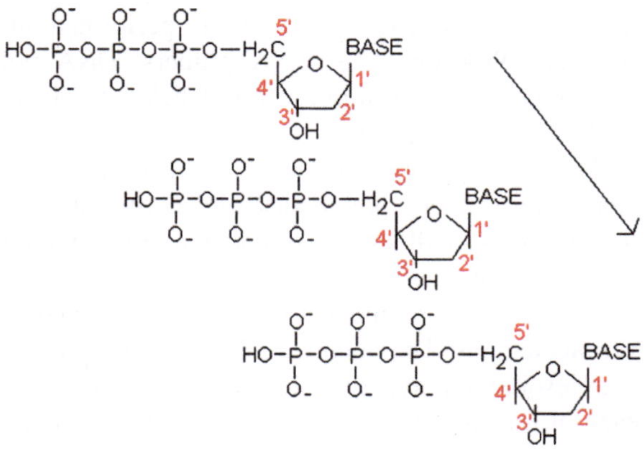

1.2 Advent of Sequencing Technologies

Knowing about the order (sequence) of nucleotides in DNA, the molecule in which the genetic information of all organisms is stored, has revolutionized biology and resulted in our better understanding of life's secrets (BBSRC Review of Next-Generation Sequencing—final version).

The first two DNA sequencing techniques, which are known as first-generation DNA sequencers, historically were developed by Fredrick Sanger (1977, University of Cambridge) and Allan Maxam and Walter Gilbert (1976–1977, Harvard University), independently. Sanger's method, which earned him a Nobel Prize in Chemistry in 1980, became popular, and in fact was the sole method for DNA sequencing for three decades, as a result of its lesser technical complexity and lesser amount of toxic chemicals used, compared to the Maxam–Gilbert method, which was based on the chemical modification of DNA and subsequent cleavage at specific bases. In the Sanger sequencing method, which is also known as "chain termination" or the "dideoxy method," modified nucleotides (fluorescently labeled dideoxynucleotides) are used in the reaction in addition to normal nucleotides; this method was gradually improved and became automated (the first automatic sequencing machine, AB370, was introduced in 1987 by Applied Biosystems), and therefore has been the method of choice for large-scale sequencing projects, e.g., whole-genome sequencing for various species, for about 30 years [4].

1.2.1 First-Generation DNA Sequencers

1.2.1.1 Sanger Sequencing Technology

In classical Sanger sequencing technology, which is sequencing by the synthesis method, the sequencing reaction is performed in the presence of the single-stranded DNA template, DNA primers, DNA polymerase, four normal DNA nucleotides, and four fluorescently labeled modified nucleotides (ddATP, ddCTP, ddGTP and ddTTP).

The DNA template is initially divided into four separate sequencing reactions containing primers, polymerase and normal nucleotides. In each reaction in the presence of a small amount of one of four modified nucleotides (which lack the 3'-OH group required for the extension), which randomly incorporates into the growing strands, terminates DNA elongation and results in DNA fragments with various lengths. The obtained DNA fragments are then separated by size through high resolution polyacrylamide gel electrophoresis (capillary electrophoresis) with each of four reactions run in one of four individual lanes (lanes A, C, G and T). DNA bands that correspond to DNA fragments with differing lengths are then visualized, using UV light or X-ray autoradiography, and the order of nucleotides can be determined according to the relative positions of DNA bands among four different lanes (Fig. 1.5).

1.2.1.2 Maxam-Gilbert Chemical Degradation DNA Sequencing Technique

The Maxam-Gilbert technique relies on the cleaving of nucleotides by chemicals and is most efficient with small nucleotide polymers (Fig. 1.6). Chemical treatment generates breaks at a small proportion of one or two of the four nucleotide bases in each of four reactions (G, A + G, C, C + T). Due to the advancements in chain termination methodology, the Maxam-Gilbert method has become redundant. It became obsolete due to its less ergonomical feasibility, and it is also considered unsafe because of the extensive use of toxic chemicals.

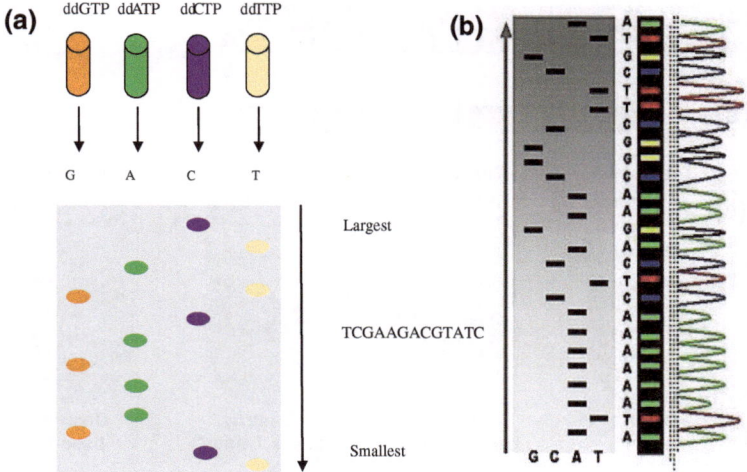

Fig. 1.5 Sanger sequencing procedure. **a** Four distinct reactions are taking place in the presence of all required materials for DNA synthesis. Besides in each separate reaction, a distinct type of fluorescently labeled dideoxy nucleotides is added which after completion DNA synthesis cycles, results in the DNA strands each of which terminated in specific dideoxy nucleotide present on that reaction. **b** After reaction completion, the content of four separate reactions is electrophoresed using high-resolution polyacrylamide gel (www.Wikipedia.org)

As a result of using less toxic chemicals and lower amounts of radioactivity than the Maxam and Gilbert method, and because of its comparative ease, the Sanger method was soon automated and was the method used in the first generation of DNA sequencers.

1.3 Some Drawbacks of the Sanger Technique

1.3.1 Short Size Fragments

The Sanger method can only be performed for DNA fragments with a fairly short length, i.e., 100–1,000 base pairs. This is due to the limitation in the power of discrimination between fragment sizes during capillary electrophoresis, which restricts the size of the DNA that can be reliably sequenced to ∼ 1,000 base pairs (for larger DNA fragments, longer gels are required). Larger sequences—for example, an entire chromosome—must first be fragmented into smaller pieces and amplified to obtain a large number of copies for each individual fragment. After performing sequencing reaction, these fragments must be reassembled to produce the original sequence.

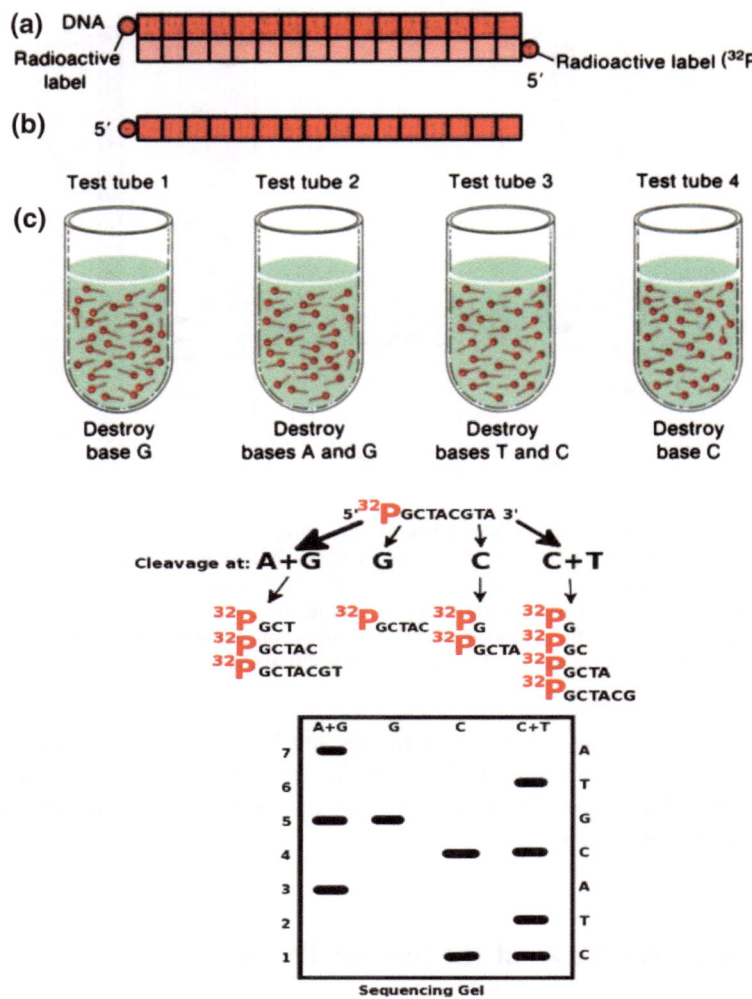

Fig. 1.6 Maxam-Gilbert chemical degradation sequencing technique. **a** Double-stranded DNA is labeled at 5′ ends. **b** Single-stranded DNA fragment is produced. **c** DNA fragments are distributed in four parallel test tubes. Each test tube is subjected to a specific base degrading chemical. The content of each tube will be electrophoresed in the next step for fragment size separation

1.3.2 Needs for Amplification and Fragment Assembly Steps

The procedure mentioned for fragmentation and amplification can be conducted by two distinct approaches: *map-based sequencing* (also known as back-to-back or hierarchical sequencing) and *shotgun sequencing*.

The map-based method is accomplished by using a large number of bacterial artificial chromosomes (BAC) (>20,000), each of which contains a large DNA fragment (approximately 100 kb), which collectively provide an overlapping series that can be physically mapped on the chromosome. After each BAC clone is amplified in a bacterial culture, it is cut into small fragments of 2–3 kb. After subcloning these fragments into a plasmid vector and amplifying them in bacterial cultures, DNA is extracted for sequencing. High coverage or over-sampling of each fragment (about eight-fold) is required to assemble the individual fragments into contigs, which are the contiguous assembled fragments. Contigs are then

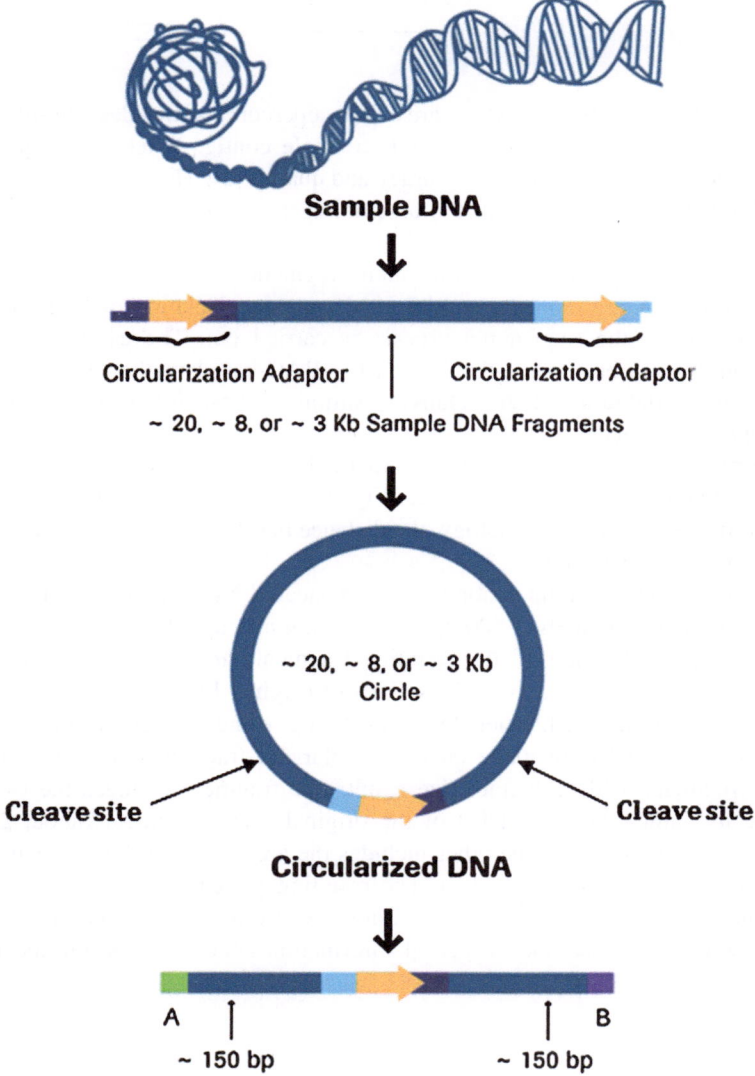

Fig. 1.7 Mate-paired library construction. Large DNA fragments are circularized using internal adapters. In the next step using endonucleases, a fragment is resulted which contains the sequences from two ends of the original sequence

Fig. 1.8 Using paired-end reads, repeats can be unambiguously aligned in complex genomes

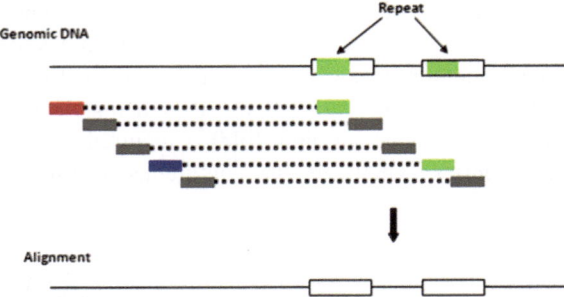

merged to form larger ones, which are called supercontigs. Besides, gap-filling and filtering must be performed to generate a single contig, which is a sequence of nucleotides with a high level of accuracy and quality [5]. The assembly algorithms and difficulties associated with the assembling process will be thoroughly dealt with in chaps. 3 and 4.

In shotgun sequencing, several small fragments that result from randomly broken up DNA are cloned into plasmids, and after isolation of each amplified DNA fragment, the sequencing process is carried out. The assembly of these fragments would be more difficult than the BAC-based method since no information is available about the relative position of these fragments on the chromosome in advance.

To overcome this problem, a new method, called "**mate-pair**" or "**paired-end**" technology, has been developed (Fig. 1.7). The two terms are almost equal; in fact, in the mate-pair technology, the distance between the two ends of the DNA fragment can be very far.

In this technology, total genomic DNA is sheared into overlapping specific size fragments (approximately 1,500 bp); then, using a "cap adaptor" and, following that, an "internal adaptor," the resulting fragments are circularized. In each of these circularized fragments, the two ends of original linear fragments have now been located beside each other. In the next step, using the activity of a nuclease enzyme, two breaks are made on the circularized fragments so that two linear DNA fragments with a distinct size result: one fragment includes the two ends (hence its name—"paired-end") of the original linear fragment (about 200 bp; 100 bp of each end), and the other includes the area between the two ends in the original linear fragment (Fig. 1.8). The resulting paired-end library can then be sequenced, and the data that results allows scaffolding of the contigs, thereby minimizing difficulties due to lack of information about the possible location of each fragment.

1.3.3 Problems with Parallelization

Automated sequencing can be performed for only 96 or 384 samples per run, and this is considered to be a limitation for parallel analyses and, consequently, massive sequencing projects such as whole genome sequencing will be too time-consuming.

1.3.4 Cost

After three decades of gradual reductions in cost, the Sanger sequencing method costs about $0.50 per kilo base, which still too high to be practical for many important research projects [6]. Since one of the final goals of sequencing techniques is to obtain the genome sequence for all individuals to fully understand genome variation, genetic susceptibility to disease and the pharmacogenomics of drug response, considerable reduction in sequencing methods is required. The National Human Genome Research Institute (NHGRI) reflected this need in 2004, and the J. Craig Venter Science Foundation announced a U.S. $500,000 prize to the group or individual who "significantly advances automated DNA sequencing…".

1.3.5 Need for Complete Automation

There are difficulties in the complete automation of sample preparation. Sanger sequencing can be performed only when high concentrations of DNA are available and, as previously mentioned, the generation of large populations of DNA fragments are carried out through bacterial cloning steps, which are tedious and time-consuming and cannot be done automatically.

References

1. Dahm, R. (2008). Discovering DNA: Friedrich Miescher and the early years of nucleic acid research. *Human Genetics, 122*(6), 565–581.
2. Pray, L. (2008). Discovery of DNA structure and function: Watson and Crick. *Nature Education, 1*(1) http://www.nature.com/scitable/topicpage/discovery-of-dna-structure-and-function-watson-397.
3. Chargaff, E. (1950). Chemical specificity of nucleic acids and mechanism of their enzymatic degradation. *Cellular and Molecular Life Sciences, 6*(6), 201–209.
4. Nowrousian, M. (2010). Next-generation sequencing techniques for eukaryotic microorganisms: Sequencing-based solutions to biological problems. *Eukaryotic Cell, 9*(9), 1300–1310.
5. Mardis, E. R. (2008). Next-generation DNA sequencing methods. *Annual Review of Genomics and Human Genetics, 9*, 387–402.
6. Shendure, J., et al. (2004). Advanced sequencing technologies: Methods and goals. *Nature Reviews Genetics, 5*(5), 335–344.

Chapter 2
Emergence of Next-Generation Sequencing

Although in the past few years the genome of several species, as well as humans, were sequenced using the automated Sanger method, the above-mentioned limitations of this method indicated a need to develop new and improved sequencing technologies to sequence the large number of human genomes and to find answers to biological problems of interest that could not be addressed before [1, 2]. For example, advances in sequencing technology would help in the development in fields such as:

1. Comparative genomics, which involves comparing the genome of distinct organisms to learn about their molecular programs.
2. Biomedical research, through which so many problems concerning the genetic basis of susceptibility to diseases, multi-factorial diseases, and cancer therapy can be investigated. The detection of different genomic and epigenomics alterations, such as single nucleotide mutations, small insertions and deletions, chromosomal rearrangements, copy number variations, and DNA methylation can be facilitated using advanced sequencing technologies [3].
3. Personal genome projects (individual genome sequencing) that impact human health, by becoming one of the major components of personalized health care by providing accurate diagnosis, prognosis and guidance for treatments [3, 4].

Subsequently, several research centers initiated the designing of new sequencing technologies not needing gels, which would allow sequencing large numbers of samples in parallel [5]. These technologies are known as Next-generation DNA sequencing (NGS) methods in which the bacterial cloning steps have been removed (in comparison with the Sanger method). Three major NGS methods that are routinely used in many laboratories today include:

1. The Roche/454 FLX (http://www.454.com)
2. The Illumina/Solexa Genome Analyzer (http://www.illumina.com)
3. The Applied Biosystems SOLiDTM System (http://marketing.appliedbio-systems.com)

A. Masoudi-Nejad et al., *Next Generation Sequencing and Sequence Assembly*,
SpringerBriefs in Systems Biology, DOI: 10.1007/978-1-4614-7726-6_2,
© The Author(s) 2013

Another three massively parallel technologies that have been introduced more recently include the Polonator (Dover/Harvard), the HeliScope Single Molecule Sequencer technology (Helicos; Cambridge, MA, USA) [6, 7], and the Ion Semiconductor (Torrent Ion Sequencing). The single molecule real time (SMRT) [Pacific Biosciences] and Nanopore Sequencing [8] are another two newly introduced technologies that are based on the sequencing of single molecules. The biochemical reaction principles of each of the above-mentioned methods will be described later.

De novo sequencing versus resequencing

There are two major types of sequencing projects in terms of application. In de novo sequencing, the genome of an organism is sequenced for the first time. In resequencing projects, the whole genome of an organism—or parts of it—is sequenced while the reference sequenced genome for the species of that organism is already available.

Sequencing depth (coverage) versus sequencing breadth

Maximum sequencing efficiency is achieved as a consequence of both depth (coverage) and uniform read distribution (breadth). Sequencing depth or coverage concerns the average number of times each base in the genome is sequenced. For example, to sequence a 3 Gb human genome with 10x coverage, 30 Gb of sequenced data is needed. Sequencing breadth refers to the percentage of the genome that is covered by sequenced reads.

2.1 454 Pyrosequencing

454 Life Science (Branford, CT, USA) was the first next-generation sequencing technology that was commercially available (in 2005) (later acquired by Roche) [7] and was also the first NGS technology to sequence a complete human genome, that of Dr. James D. Watson [9]. This technique was first successfully validated by sequencing of the entire 580,069 bp of the *Mycoplasma genitalia* genome at 96 % coverage and 99.96 % accuracy in a single run. This method is based on using an alternative sequencing technique called "pyrosequencing," which was first introduced by PålNyrén and Mostafa Ronaghi at Stockholm's Royal Institute of Technology in 1996. The maximum length of sequenced reads by the 454 System was 450 bp at first and has now been increased to about 700 bp. (Note that most of the road's length presented here for discussed platforms is subjected to rapid changes and are appropriate just for comparison between different platforms.) In pyrosequencing, which uses the "sequencing by synthesis" concept (it differs from the Sanger sequencing method in that it depends on the detection of pyrophosphate release on nucleotide incorporation rather than chain termination with dideoxy-nucleotides) [10], the incorporation of each of four nucleotides which are added in a certain order in each step results in the release of pyrophosphate, which in the presence of adenosine $5'$ phosphosulfate can be converted to ATP by enzyme ATP

sulfurylase. This ATP in turn acts as a substrate for a chemiluminescent enzyme, luciferase, which converts luciferine to oxyluciferine and visible light that can be detected by a charged coupled device (CCD) camera for further analyses. The amount of generated light is proportional to the number of incorporating nucleotides. Since dATP can act directly as a substrate for Luciferase and subsequently generate light without necessarily needing for incorporation of dATP, in order to prevent any bias, dATPαS, which can precisely incorporate DNA strand while not a substrate for luciferase enzyme, is used instead of dATP. In each step, unincorporated nucleotides are degraded, using an enzyme called apyrase to exclude the possibility of incorporation of nucleotides remaining from the previous steps. The process can be continued with another step of nucleotides additions [11].

In the 454 approaches, genomic DNA is first fragmented into smaller pieces and short adaptors are then ligated to blunt ends (3′ and 5′ ends) of the single-stranded fragments (Fig. 2.1a). These adaptors flanked fragments are then mixed with small 28-μm streptavidin-coated beads whose surfaces have been covered with short sequences complementary to one of the ligated adaptors. Fragments are then hybridized to their corresponding beads in such a way that each bead carries a unique fragment. Fragment amplification must be performed for intensifying the light signal that is required for precise detection of added bases by CCD camera [5]. Amplification of DNA fragments are performed through emulsion PCR in which all required reagents are provided in droplets of water in oil mixture which acts as a micro-reactor for every single bead; consequently, it guarantees that every bead is covered by single-type strands at the end of the amplification step. After breaking the emulsions, the beads whose surfaces have been now covered with about one million copies of each amplified fragment are first pre-incubated with the DNA polymerase enzyme and then loaded onto a PicoTiterPlate device so that each single bead is deposited in a well with the dimensions that can contain only one bead [6]. The wells are then filled with smaller beads (1-μm) that carry immobilized enzymes required for pyrosequencing (sulfurylase and leuciferase) and also help beads to deposit in the wells gravitationally [12]. PicoTiterPlates are centrifuged to ensure that the enzymes are in close contact with the DNA beads [13]. The loaded plates are placed in a 454 FLX instrument in which sequencing reagents (dNTPs and buffers) are delivered to the wells of the plate. The pyrosequencing technique is initiated with the sequential addition of four nucleotides, as previously explained (Fig. 2.1b).

When a nucleotide is added to the growing chain, a signal is recorded by a CCD camera as a result of the generation of light. As mentioned, the signal strength is proportional to the number of nucleotides, and in the case of homopolymers (the stretches of only one nucleotide type), the signal intensity is higher than that of a single nucleotide. However, signal strength can be precisely detected only for less than 10 consecutive nucleotides; after that, the signal declines rapidly [7]. As a result of this drawback, the insertion and deletion error rates are the most frequently observed errors in this method. The substitution error rate of this method (10^{-3}– 10^{-4}) is higher than the rates for the traditional Sanger method. However,

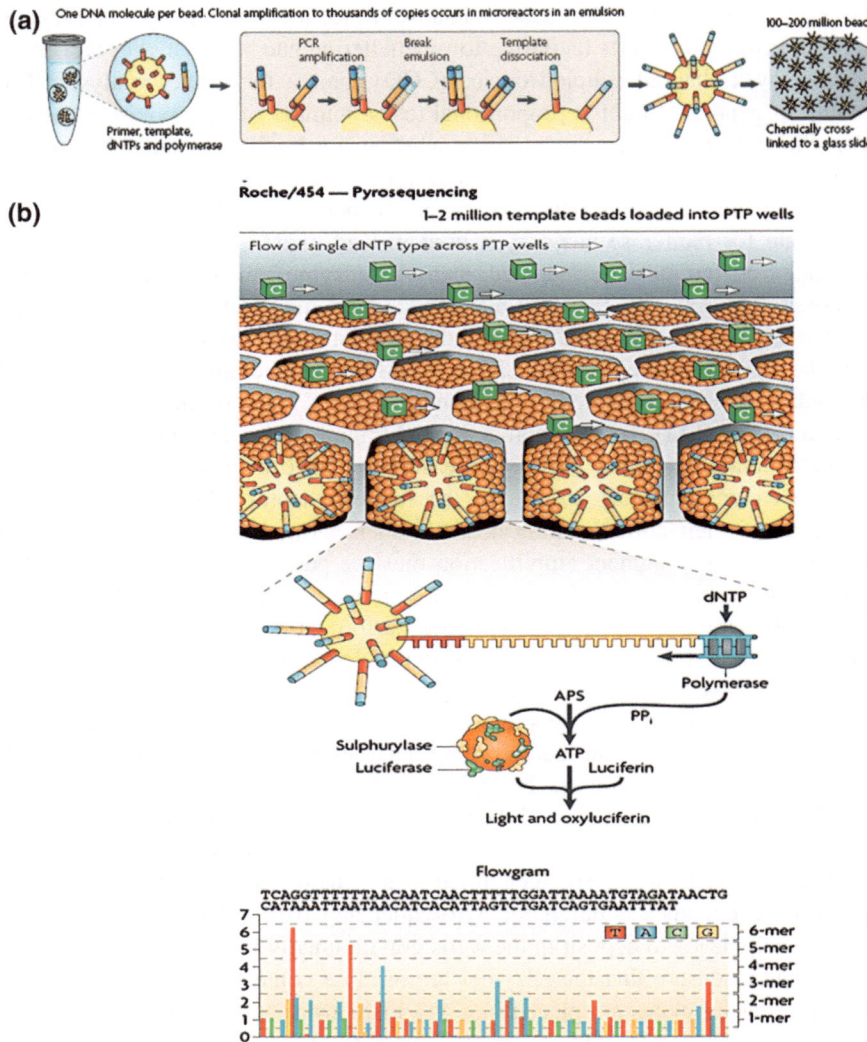

Fig. 2.1 A schematic representation of 454 pyrosequencing technology. **a** Construction of adaptor flanked DNA fragment, DNA amplification on the beads' surface and formation of "clone beads." **b** Deposition of beads into "picotiter plate" wells. In the pyrosequencing method, incorporation of each nucleotide is determined by the detection of light generated by the activity of luciferase enzyme in the presence of ATP and luciferine [5]

the background error rate of the Sanger method that arises from *in vitro* amplification steps that introduce the error in the sample before performing sequencing, is higher [12].

2.2 Illumina (Solexa) Genome Analyzer

The Genome Analyzer first introduced by Solexa in 2006 (San Diego, CA, USA) and then further developed by Illumina. At first the resulting reads were very short (36 bp or less) compared to Sanger methods. Since then, many technical improvements have been introduced to this method, resulting in increased read length to up to 100 bp [12]. Since millions of reads can be generated in parallel (simultaneously), reassembling of these reads using efficient algorithms results in useful outputs.

In this method, a DNA library can be constructed by any method that generates adaptor-flanked fragments up to several hundred base pairs in length (Fig. 2.2a) [6]. Template amplification is performed with a method called "bridge PCR," in which both forward and reverse primers that are complementary to the adaptors' sequences are attached to a solid surface called "flow cell," containing eight independent channels that allow performing PCR amplification on its surface. The name "bridge PCR" refers to the fact that during the annealing step, the extension product from one bound primer forms a bridge to the other bound primer and its complementary strand is synthesized. Since each initial single fragment has been tethered to a distinct channel on the surface, the amplified fragments form clusters, each of which contains approximately one million copies of the initial fragment that would be sufficient for sequencing reaction (Fig. 2.2b). Moreover, each cluster contains both forward as well as reverse strands of the original sequences, but to have homogenous populations of strands in each cluster, which can be sequenced precisely without the interference of the complementary strand, one of the strands must be removed before initiating the sequencing process [12]. The sequencer adopts the technology of sequencing by synthesis. In this method, four reversible blocked nucleotides (3'-OH is chemically blocked), which then transiently block the extension of nucleotides, are added to the reaction. Following the incorporation of each nucleotide, the imaging step is carried out. After the acquisition of images in each cycle, the 3'-OH blocking group is chemically removed and another cycle can subsequently be initiated. The steps of adding nucleotides and other PCR reagents and capturing image can be continued for a specific number of times and ultimately generate 100 bp reads.

The first Illumina sequencer, GA, as already described, was introduced in 2006. The performance of GA was improved from 1G/run to 85G/run in the GAIIX series, which was released in 2009. In early 2010, Illumina launched HiSeq 2000, which adopts the same sequencing strategy with GA. Its initial output was 200G per run, and improved to 600G per run currently, which could be finished in 8 days. In 2011, MiSeq, a bench-top Illumina sequencer, was introduced; it shared most technologies with HiSeq. MiSeq is especially convenient for amplicon and bacterial sample sequencing [14].

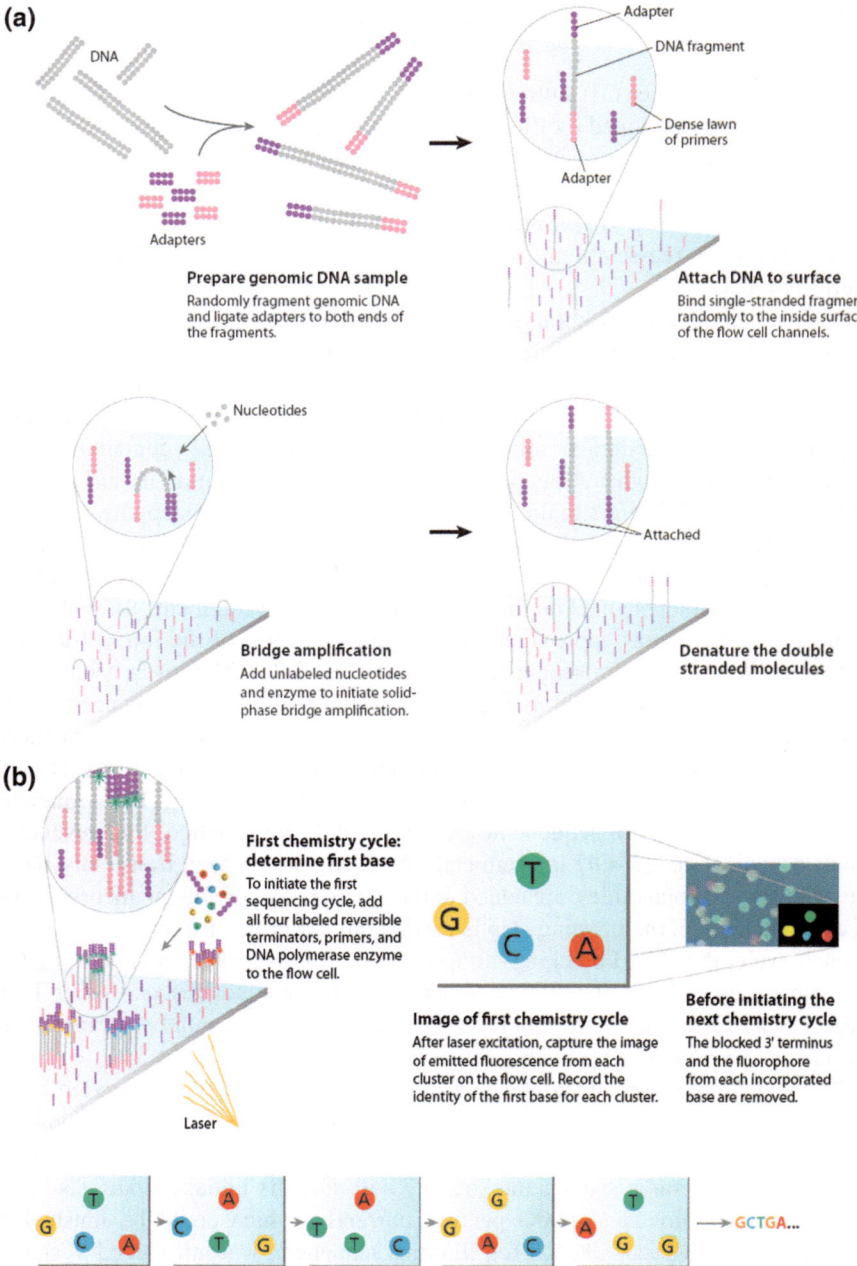

Fig. 2.2 Illumina (solexa) sequencing by synthesis technology. **a** Clonal amplification of adaptor flanked fragments is performed by bridge PCR and then single-stranded DNA is produced for sequencing step. **b** Base detection is conducted by the addition of four labeled reversible blocked nucleotides, primers and a DNA polymerase enzyme. After each nucleotide incorporation, the image of emitted fluorescence is captured and then the 3′ blocked nucleotide and the fluorescent dye are removed to permit initiating another cycle of nucleotide detection [7]

2.3 Applied Biosystems SOLiD Sequencing

The SOLiD system, which became commercially available in late 2007, was developed by J.S. and colleagues in collaboration with McKernan and colleagues at Agencourt Personal Genomics, later acquired by Applied Biosystems and now part of Life Technologies (Carlsbad, CA, USA). In this method, the library of adaptor-flanked DNA fragments is constructed similarly to the previously mentioned NGS methods. Clonal amplification is performed, as mentioned earlier, for the 454 pyrosequencing platform through emulsion PCR using small magnetic beads (1-μm in diameter) to which DNA fragments can be bound, so that each bead carries an individual fragment. After amplification of DNA fragments, "bead clones" that have now been covered by nucleic acid species, are selectively isolated and are tethered to a solid surface—a "flow cell," via 3′ modification of DNA strands. This flow cell is basically a microscope slide that can be serially exposed to liquids. A universal primer, which is complementary to the common adaptor in each fragment, is annealed to the DNA fragment. In the SOLiD method, unlike other NGS technologies, which base detection of DNA fragments, is performed through polymerase reaction; this is achieved by sequencing by ligation. A library of fluorescently labeled octamers (four distinct fluorescent dyes) is constructed with the following composition: each octamer has a specific dinucleotide (out of 16 possible combinations), which in different versions of ABI-SOLiD sequencing chemistry is different (for example, in ABI SOLid v.2.0 positions 1 and 2 and in the original version, bases 4 and 5 are utilized). The rest of the bases have completely random (degenerate) composition. Each octamer also holds one of four fluorescent dyes—most often at position 6, which correlates with the composition of dinucleotide in that octamer. The sequencing process can be conducted as follows:

(1) A universal primer with "n" nucleotides anneals to each DNA fragment, and then a mixture of octamers with specific dinucleotide composition is added and one whose specific dinucleotide is matched with a template then can be hybridized, which will be followed by the process of ligation in the presence of enzyme ligase.
(2) In this step, an image is taken (the specific fluor is detected, which corresponds to the specific dinucleotide composition; in fact, each fluorescent dye is in correlation with four dinucleotide combinations—see Fig. 2.4). The unextended fragments are capped in the presence of the same mixture of nonfluorescent dinucleotides.
(3) In the next step, the last three bases of octamer (i.e., the bases 6, 7 and 8) along with the fluorescent dye are chemically removed, so five bases of octamer are left behind and the above steps are repeated for base 6 and 7. This step is typically repeated 10 times [12] and at the end, the fluorescent colors, corresponding to 10 dinucleotides (1/2, 6/7, 11/12, 16/17, 21/22, 26/27, 31/32, 36/37, 41/42 and 46/47) is determined.

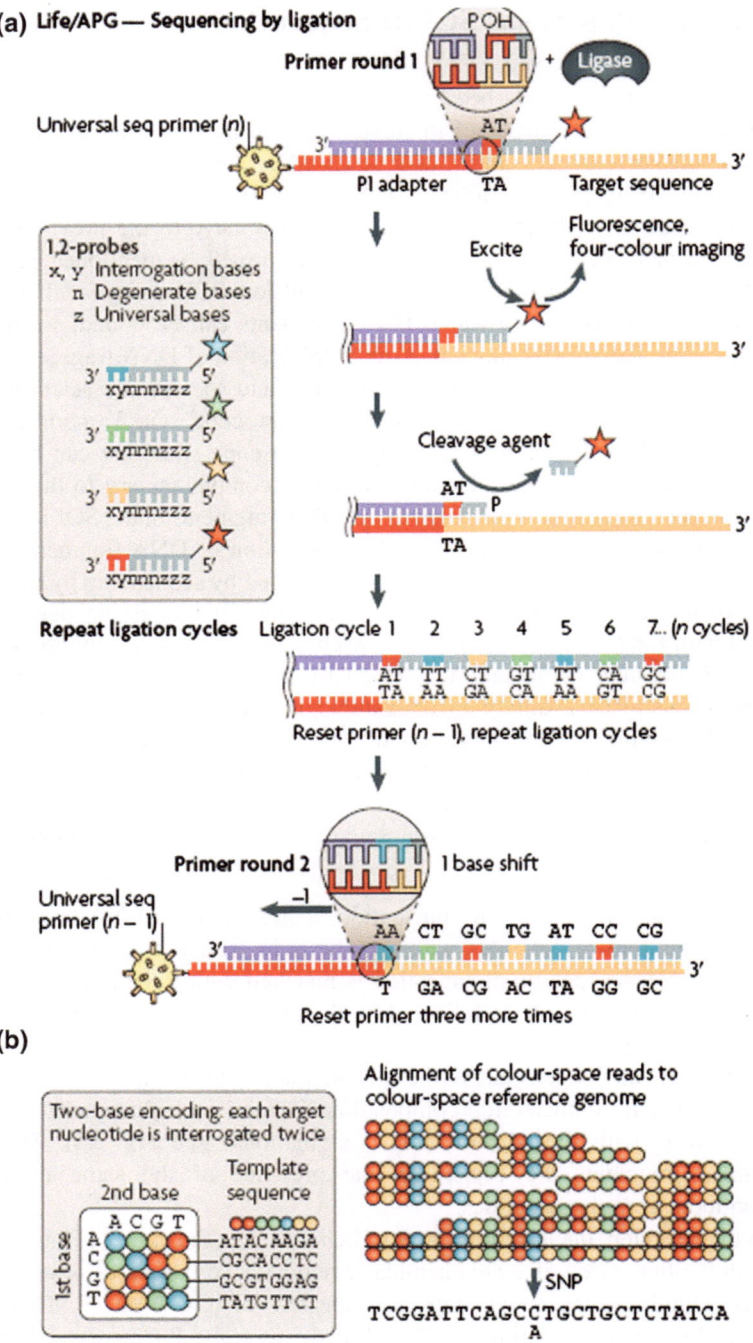

(a) Life/APG — Sequencing by ligation

Primer round 1

Universal seq primer (*n*)

P1 adapter   TA   Target sequence

Ligase

1,2-probes
x, y Interrogation bases
n   Degenerate bases
z   Universal bases

3′ xynnnzzz 5′
3′ xynnnzzz 5′
3′ xynnnzzz 5′
3′ xynnnzzz 5′

Excite   Fluorescence, four-colour imaging

Cleavage agent

Repeat ligation cycles   Ligation cycle 1   2   3   4   5   6   7... (*n* cycles)

AT TT CT GT TT CA GC
TA AA GA CA AA GT CG

Reset primer (*n* − 1), repeat ligation cycles

Primer round 2   1 base shift

Universal seq primer (*n* − 1)   −1

AA CT GC TG AT CC CG
T   GA CG AC TA GG GC

Reset primer three more times

(b)

Two-base encoding: each target nucleotide is interrogated twice

                Template
2nd base   sequence

        A C G T
    A   ATAC AAGA
    C   CGCA CCTC
    G   GCGT GGAG
    T   TATG TT CT

Alignment of colour-space reads to colour-space reference genome

SNP

TCGGATTCAGCCTGCTGCTCTATCA
                A

◄ **Fig. 2.3** Schematic representation of the steps involved in AB-SOLiD sequencing. **a** After initial primer (n-nucleotides) annealing, seven cycles of oligonucleotide hybridization and ligation, imaging and cutting the 5′ three nucleotides along with fluorescent dye is performed. Then the initial primer and extended strand are removed, which is followed by annealing a new primer with n-1 nucleotides (primer reset). Following that, all the above-mentioned steps are repeated. Repeating primer resets four times results in 35-base generation. **b** Dinucleotide encoding by four fluorescent dyes and decoding of each base using two dinucleotides. Alignment of the color-space reads of the particular sequence with color-space reads of the reference genome will result in SNP identification [1, 7]

(4) In the next step, the initial primer and all ligated portions are melted and discarded, and another round of primer annealing, "primer reset," is started with a primer with "n-1" nucleotides via repeating steps 1–3, and this time the position of nucleotides 0/1, 5/6, 10/11, 15/16, 20/21, 25/26, 30/31, 35/36, 40/41 and 45/46 is detected. After five primer reset cycles, approximately 50 contiguous nucleotides have been generated. In each of which nucleotides have been detected by two separate dinucleotides (Fig. 2.3a).

Since there are 16 possible combinations of dinucleotides (4^2) and only four fluorescent dyes (Fig. 2.3b), nucleotide identity detection would not be possible only from data color obtained from the incorporation of dinucleotides. The dinucleotide composition of the first dinucleotide, i.e., nucleotides 0/1, is readily deduced from its fluorescent signal because the first base of newly synthesized strand corresponds to the last base of the universal primer, whose composition is obvious and will help decode the composition of other dinucleotides from their corresponding fluorescent signals [12]. An alternative procedure to primer reset after the first round discussed above is to use octamers with different positions of dinucleotides. For example, in the second round, different positions of octamer mixture (e.g., bases 4 and 5) can be correlated with the identity of fluorescent dye [8].

2.4 Ion Semiconductor (Ion Torrent Sequencing)

Ion Torrent sequencing technology uses simple sequencing chemistry in which no enzymatic reaction, no fluorescence, no optic, and no light are used to determine the sequence of a DNA fragment (Life technologies). This technology, which is one of the least expensive methods, has been introduced to research and clinical laboratories as a personal genomic machine. The biochemical basis of this system is very simple and involves the release of hydrogen ion following the incorporation of a nucleotide into a strand of DNA by DNA polymerase (Fig. 2.4a). To carry out this biochemical process in a highly parallel way, Ion Torrent uses a high-density array of micro-machined wells that is provided on the Ion Torrent proprietary microchips. There is an Ion-sensitive field effect transistor (ISFET) below the wells that detects the change in pH as a result of hydrogen release. This change is

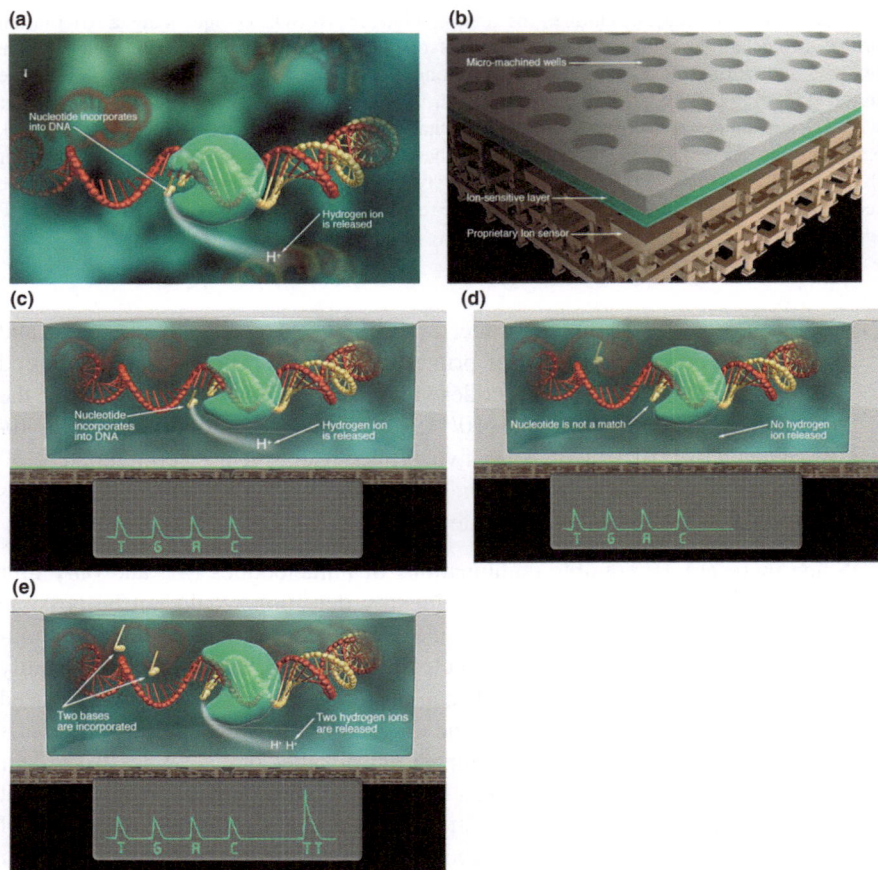

Fig. 2.4 Schematic work flow of Ion Torrent sequencing technology. **a** Release of a hydrogen ion as a byproduct following the incorporation of a nucleotide. **b** High-density array of micro-machined wells. A different DNA template is held in each well. Beneath the wells is an ion-sensitive layer and beneath that a proprietary Ion sensor. **c** Following the incorporation of the added nucleotide and release of hydrogen ion, change of the pH (chemical signal) is translated into digital information. **d** In the case of no incorporation, no voltage change is recorded. **e** If there are two identical bases on the DNA strand, the voltage will be double (taken from Life Technologies™)

recorded as a potential change (ΔV) by an ion sensor layer that indicates the nucleotide incorporation event [15] (Fig. 2.4b).

Calling the bases by this method goes like this: Following the addition of a certain base to a DNA template ("A," for example), if it is incorporated in to the DNA strand, a hydrogen ion will be released and consequently the charge of that ion will change the pH of the solution that can be detected by the ion sensor (Fig. 2.4c). The chip will be flooded by introducing one nucleotide after another. If no incorporation occurs following the addition of a nucleotide, no voltage change is recorded, and as a result no base will be called. In the case of the presence of two consecutive identical bases on the DNA strand, the voltage will be doubled, and as a result, two identical bases will be recorded. The size of DNA fragments that can be sequenced by this method was first 100 bp and now has been improved to 400 bp.

2.5 Polonator Technology

Polony sequencing technology was first introduced by Dr. George Church's group at the Harvard Medical School. Unlike other sequencing technologies, it has open-source software and free downloadable protocols. The polony sequencing method is begun by paired-end-tag library construction (Fig. 2.5). The Target DNA sequence is randomly sheared, and then the fragments—about 1 kb in size—are selected. After making the ends of these fragments blunt and A-tailing, in which an A is added to the $3'$ ends of fragments, the fragments are circularized, using 30 bp-synthesized oligonucleotides (T30) with two outward facing recognition sites for a type II restriction enzyme (Mme1); then amplification will occur, using rolling circle replication. In the next step, the amplified circularized DNA is subjected to MmeI, which cuts at a distance of 17–18 bp after detecting the recognition site, and this results in the generation of a fragment of about 70 bp from which 30 bp belongs to T30 and the rest belongs to two 17–18 bp flanking regions or tags. The resulting fragments then end repaired and two emulsion-PCR primers will be attached to their $3'$ and $5'$ ends, resulting in the production of a 135 bp fragment that is then subjected to amplification. These 135 bp fragments construct a paired-insert library. Emulsion PCR is then performed on these fragments, using strep-tavidin-coated beads that have a complementary oligonucleotide of forward primer in each 135 bp fragment. Following that, the resulting emulsion droplets are broken. As a result of inequality in the distribution of the templates on the surface of the beads, the emulsion PCR yields empty, clonal and non-clonal beads [8], and an enrichment procedure must consequently be conducted to produce a dense population of clonal beads. The beads are then immobilized on the surface of the flow-cell for sequencing.

The following steps are carried out during polonator sequencing: after inserting the flow cell in the polonator instrument, each flow cell is subjected to a stream of a mixture of anchored primers where they can be hybridized to the 3 or 5 ends of the two 18 bp flanking oligonucleotides (a total of four possible positions are available).

Following that, a completely degenerate mixture of fluorescently labeled nona-mers (in each of which the identity of one position corresponds to the identity of the fluorescent dye that is bound to it: $5'$ Cy5-NNNNNNNNT, $5'$ Cy3-NNNNNNNNA, $5'$TexasRed-NNNNNNNC, $5'$ 6FAM-NNNNNNNNG) and T4 DNA ligase flow into the cell. A fluorescent signal is generated when ligation selectively occurs between the anchored primer and corresponding nonamer. In the following step, the

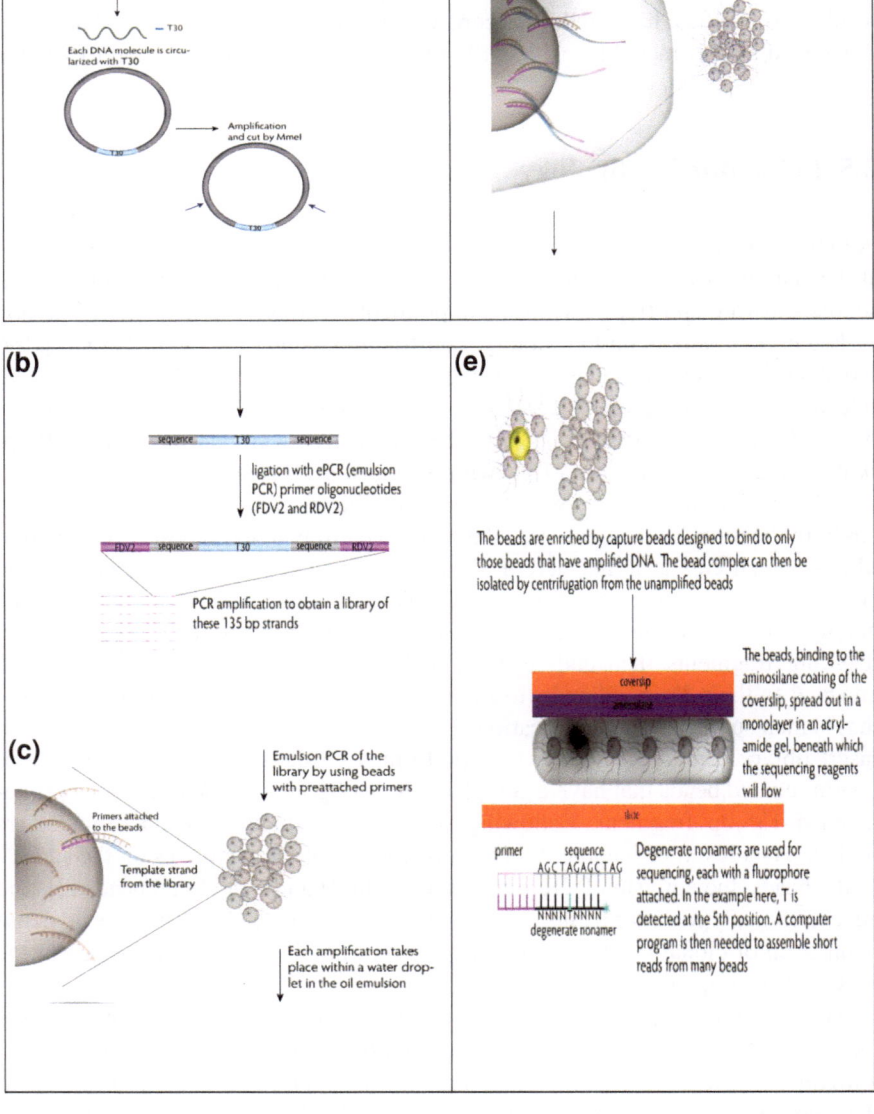

Fig. 2.5 Steps for Polony sequencing method. **a** Paired genomic tag construction using universal linker (T30) for circularizing DNA fragment and MmeI for making a cut 17–18 bp outside T30. **b** Ligation of two emulsion PCR primers at the ends of the resulted fragment. **c** In the next step, emulsion PCR is performed, resulting in empty, clonal and non-clonal beads. **d** The enrichment step, which involves bringing together the clonal beads. **e** Conducting sequencing using continuous cycles of occurring ligation process between anchored primers and fully degenerate fluorescently labeled nonamers in each of which the identity of only one base is known and a fluorescent label has been attached to each nonamer according to that base

imaging step is conducted, and after that the primer-nonamer complex and ligase are removed and then another cycle of sequencing can be begun with a new mixture of nonamers in which the query position is one base shifted further into the 17–18 bp tags (5′Cy5-NNNNNNNTN, 5′Cy3-NNNNNNNAN, 5′TexasRed-NNNNNNNCN, 5′ 6FAM- NNNNNNNGN) and new primers. Using DNA ligase, the query position can be precisely determined when the distance between it and the ligation site is 7 bases in 5′ to 3′ direction and 6 bases in 3′ to 5′ direction. As a result, totally 26 bp from two 17–18 bp tags (13 bp from each) can be sequenced with a 4–5 bp gap within each tag.

2.6 Heliscope (Single Molecule Sequencing)

Heliscope, which was one of the first techniques for sequencing a single DNA molecule, was introduced by Braslavsky et al. in 2003 and licensed by Helicos biosciences in 2007. Using this method, large research projects can be conducted in less time, and fewer errors and lower expenses since there is no need for DNA sample amplification [16]. This method, which is carried out on single DNA molecules, is conducted on an instrument called the Heliscope sequencer and uses the sequencing by synthesis technique. The DNA sample is first sheared into small pieces (100–200 bp) and then becomes adaptor-flanked. This adaptor is usually a poly-A tail, and adaptor-flanked fragments can consequently be tethered to a surface of flow cells on which poly-T oligonucleotides have been bound (Fig. 2.6). After the attachment of fragments, a mixture of a single-labeled nucleotide and polymerase are streamed into the surface and the polymerase will add nucleotide wherever it is complementary to the first positions of the attached fragments. Then the image is taken from all bound fragments that contain incorporated labeled nucleotides. The next cycle of nucleotide incorporation can be performed when the label attached to incorporated nucleotides is cleaved off (this is somewhat similar to the Illumina technology, in which reversible blocked nucleotides were utilized). Nucleotide incorporation can be continued up to 25–45 bases.

2.7 Latest Developments in Next-Generation Sequencing Methods

Since introducing the Helicos system, many attempts have been made to generate efficient sequencing technologies relying on single molecules, excluding the need for an amplification step. The two most promising such technologies are nanopore DNA sequencing and Pacific single molecule real time (SMRT) DNA sequencing.

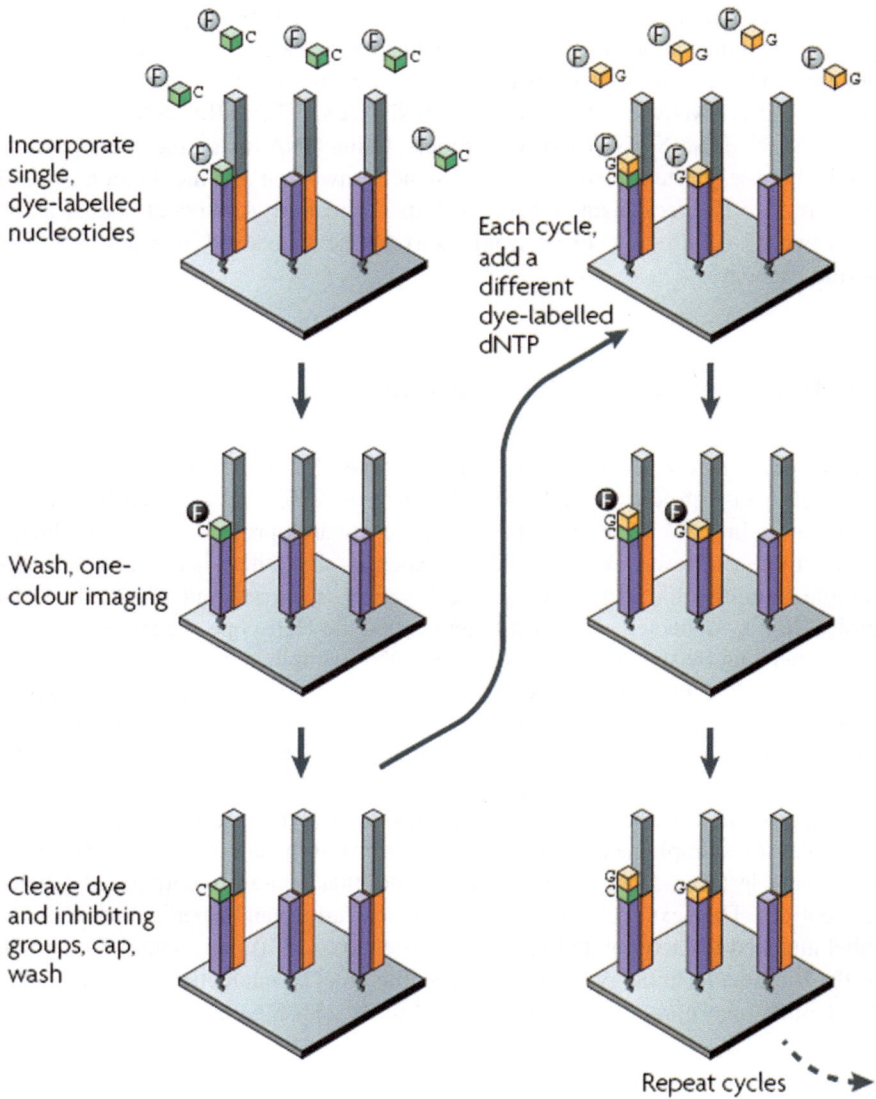

Incorporate single, dye-labelled nucleotides

Each cycle, add a different dye-labelled dNTP

Wash, one-colour imaging

Cleave dye and inhibiting groups, cap, wash

Repeat cycles

Fig. 2.6 Schematic of Helicos sequencing method. After shearing a single DNA molecule into small pieces and poly-A tailing (shown in *orange*), these fragments are bound to the surface of the flow cell using poly-T already attached to the surface (shown in *dark blue*). Once polymerase and labeled single base flow into the surface, nucleotide incorporation occurs wherever complementarities exist between the added base and the first position of attached DNA fragments. Unincorporated bases are washed from the flow cell. Then a camera scans the entire surface to realize on which fragments the new bases have been added. In the next step, the labeling dye and the inhibitor attached to incorporated bases are cleaved off and consequently another cycle of introducing new labeled nucleotide and polymerase can be begun [1]

2.7.1 Nanopore Sequencing

A few nanopore sequencing methods have been introduced; in all of them, the sequence of a DNA fragment can be identified based on the translocation of a single-stranded DNA molecule through a thin membrane. One of the main features of nanopore techniques is the ability to thread extremely long DNA molecules. Moreover, solid state sub-5 nm pores can be parallelized in the form of a very condensed array [17]. Two nanopore sequencing techniques are described here:

The technique that was introduced by Oxford Nanopore Technologies seems to be able to achieve the US National Institutes of Health's goal of decreasing the cost for sequencing of the entire human genome to $1,000 in the near future. In this sequencing platform, there is no need for any DNA fragment shearing, amplification step, fluorescently labeled nucleotides and optical instrumentation for detecting fluorescent labels. This technology has its origin in threading a single-stranded DNA molecule through a nanoscale protein pore (staphylococcal α-haemolysin) created in a membrane under an applied potential (Fig. 2.7). Passing a DNA strand through the pore results in fluctuation of the ionic current. The translocation of each base through the pore causes a decrease in current intensification which is specific for each kind of base. In fact, while each of the four bases passes through the pore, a different amount of current can translocate, and this is the key feature in base identification. However, there is difficulty in the base registration step as a result of the high speed with which DNA strand passage takes place, which in turn results in the current resolution that is essential for the precise detection of bases. This problem may be overcome by recent works that reduce the speed of DNA translocation through the pore [18].

In another nanopore sequencing technique proposed by McNally, the target DNA is first subjected to a biochemical preparation step in which each base of the sequence is converted into a form that can easily be read using a solid state nanopore. In fact, each of the four bases (A, C, G, and T) in the target DNA is converted to a predefined sequence of oligonucleotides, which is hybridized with a molecular beacon that carries a specific fluorophore (McNally). Molecular beacons are oligonucleotide probes that can report the presence of specific nucleic acids in homogenous solutions. Molecular beacons are hairpin- shaped molecules with an internally quenched fluorophore whose fluorescence is restored when they bind to a target nucleic acid [19]. For a two-color readout (i.e., two types of fluorophores), the four sequences are combinations of two predefined unique sequences, bit "0" and bit "1," such that an A would be "1 1,", a G would be "1 0," a T would be "0 1" and finally a C would be "0 0" (Fig. 2.8a). Two types of molecular beacons, carrying two types of fluorophores, hybridize specifically to the "0" and "1" sequences. The converted DNA and hybridized molecular beacons are electrophoretically passed through a solid-state pore where the beacons are sequentially removed. Each time a beacon is removed, a new fluorophore is lighted, which results into a burst of photons recorded at the location of the pore (Fig. 2.8b). This method allows wide field imaging and spatially fixed pores that make possible the simultaneous detection of

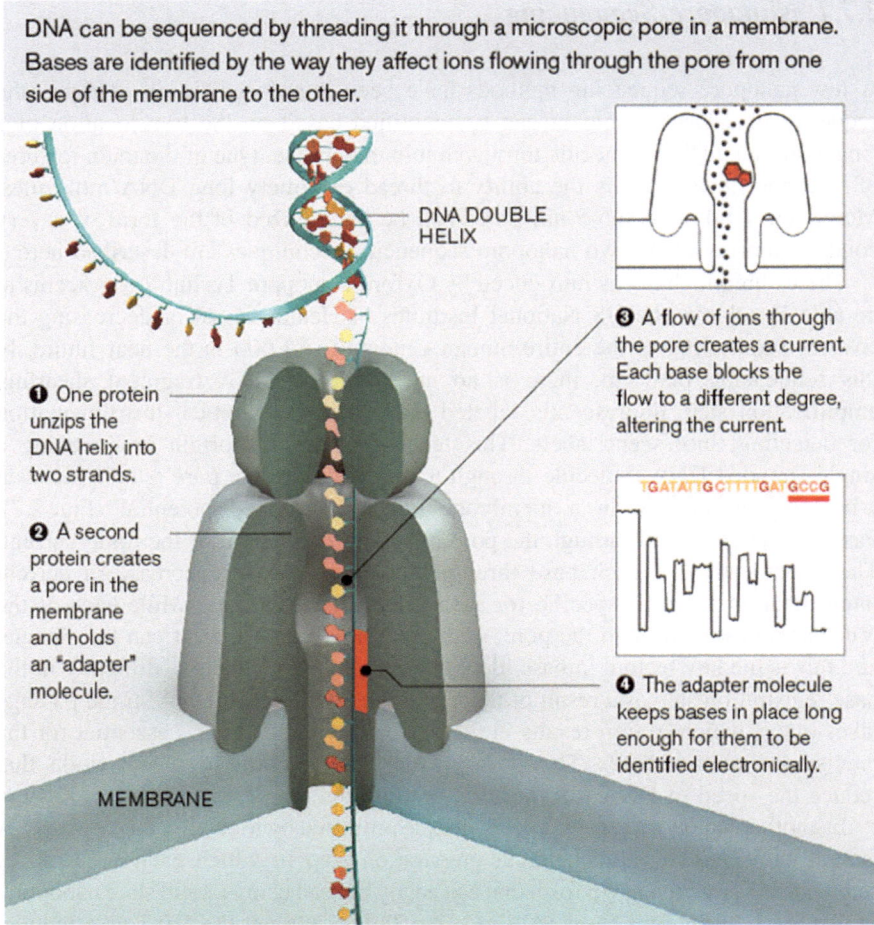

DNA can be sequenced by threading it through a microscopic pore in a membrane. Bases are identified by the way they affect ions flowing through the pore from one side of the membrane to the other.

DNA DOUBLE HELIX

❸ A flow of ions through the pore creates a current. Each base blocks the flow to a different degree, altering the current.

❶ One protein unzips the DNA helix into two strands.

❷ A second protein creates a pore in the membrane and holds an "adapter" molecule.

TGATATTGCTTTTGATGCCG

❹ The adapter molecule keeps bases in place long enough for them to be identified electronically.

MEMBRANE

Fig. 2.7 Schematic representation of nanopore sequencing system. (1) The upper protein is used to make the DNA molecule single stranded. (2) The second protein forms a nanopore in a membrane. It also contains an adaptor molecule. (3) Each base obstructs the flow to a different degree. (4) The adaptor is basically used to reduce the speed of passing DNA through the pore, which is necessary for the exact identification of the DNA strand base composition (picture from MIT's Technology Review)

several pores using a special camera, Electron multiplying charged coupled device (EM-CCD) [17].

2.7.2 Single Molecule Real Time DNA Sequencing

This technique was introduced by Pacific Biosciences in 2009. It is based on the observation of the performance of polymerase during DNA synthesis. On this

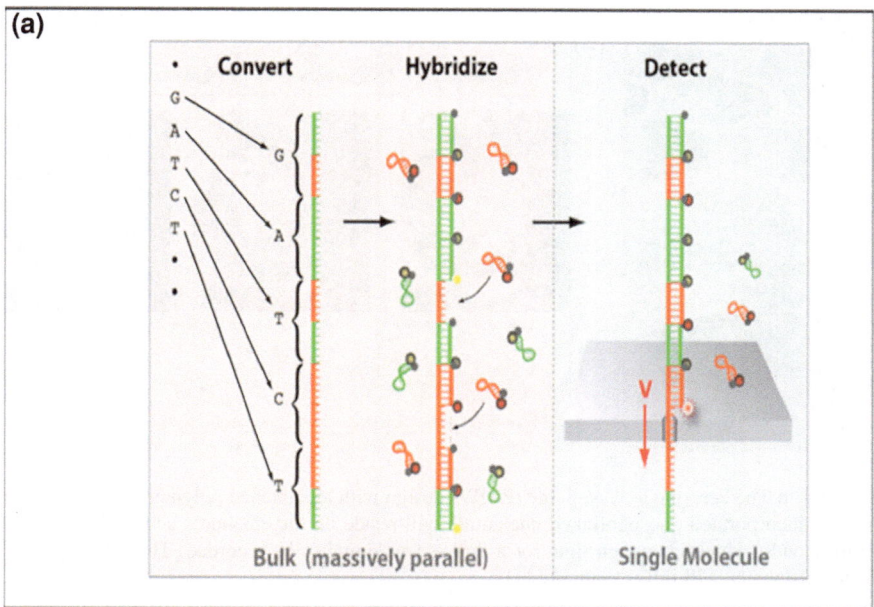

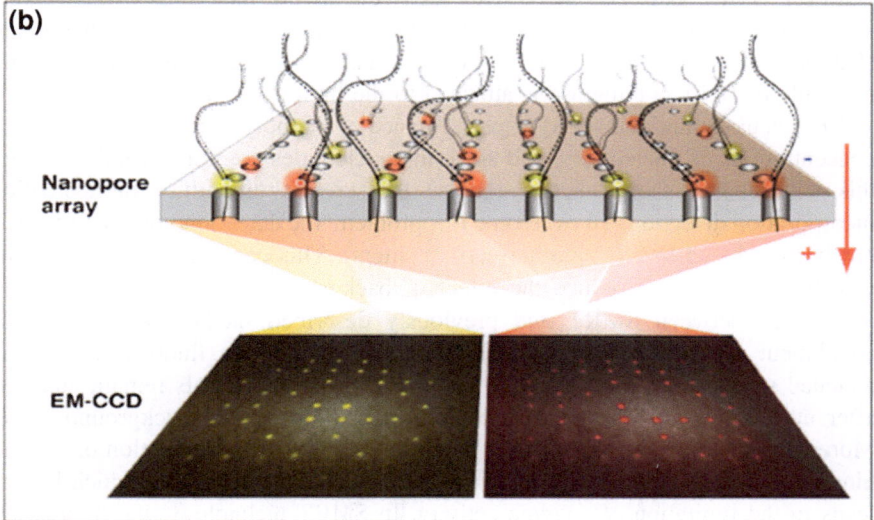

Fig. 2.8 Another version of nanopore sequencing methods introduced by McNally. **a** The biochemical preparation step involving conversion of each base of the sequence into an oligonucleotide that can be hybridized with a molecular beacon. **b** The threading of the beacon hybridized oligonucleotides through a nanopore makes it possible to detect optical signals. The signals are projected onto a wide field imaging screen that is very useful in the simultaneous detection of several pores using an EM-CCD camera [17]

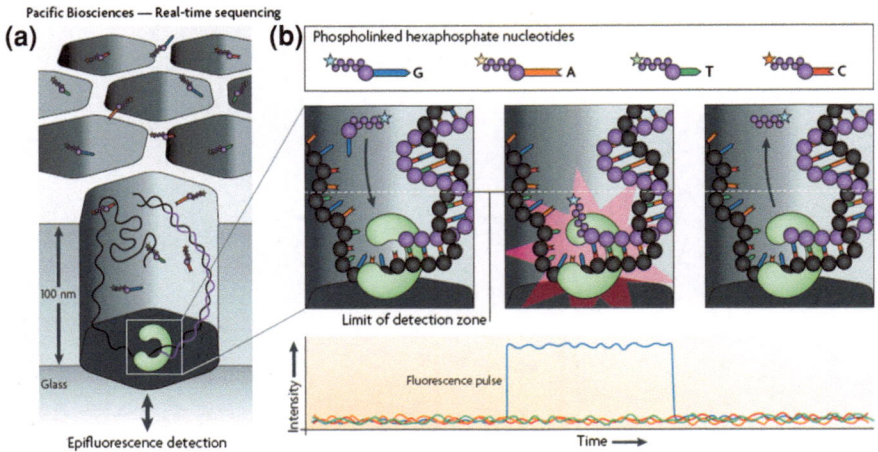

Fig. 2.9 a The zero-mode waveguide (ZMW) design with an attached polymerase at its bottom. **b** Each incorporated phospholinked nucleotide will reside on the enzyme's active site for a few milliseconds, which is enough time for a fluorescent signal to be recorded. The released labled pentaphosphates will diffuse quickly [1]

platform, SMRT cells are used, with each cell having thousands of zero-mode waveguides (ZMWs), which are holes in a surface that acts as a nanoscale chamber. In each ZMW (which is tens of nanometers in diameter), a single molecule of DNA polymerase is attached to the bottom surface (Fig. 2.9a).

The accuracy and the speed of performance of the polymerase depend on high concentrations of nucleotides, and since the nucleotides are fluorescently labeled, this will lead to the background noise that creates difficulties in nucleotide incorporation detection. To overcome this problem, the detection volume in SMRT has been reduced to 20 zeptoliters (10^{-21} liters). This considerable reduction of detection volume can reduce the effect of background noise. One of the main differences between SMRT and previously described methods is the site of attachment of the fluorescent label. In other systems, the fluorescent label is attached to the base in nucleotides and consequently the labels remain attached after nucleotide incorporation, which leads to an increase in background noise. Moreover, incorporation of multiple bases will also lead to the creation of a steric hindrance as a result of the physical bulk of several dye molecules, which in turn leads to the limitation of enzyme activity. In SMRT technology, the fluorescent label is attached to the phosphate chain, and as a result of nucleotide incorporation the pentaphosphate-label couple will be removed from the nucleotides and will diffuse out of the reaction volume (Fig. 2.9b) [20], (Pacific Biosciences, 2009. Single Molecule Real Time (SMRT^{TM}) DNA Sequencing) [1]).

2.8   Comparison of Available Next-Generation Sequencing Techniques

Already available next-generation sequencing methods are distinguished from each other based on various characteristics including maximum read length, number of reads in each run (degree of parallelization), run time, and many others. Table 2.1 compares the performances of various next-generation sequencing instruments [9]. Note that most of the information provided here is subject to rapid changes and it is suitable just for general comparison between techniques.

2.9   DNA Sequencing Costs

For many years, the National Human Genome Research Institute (NHGRI) has tracked the costs associated with DNA sequencing performed at the sequencing centers funded by the Institute. This information serves as an important benchmark for assessing improvements in DNA sequencing technologies and for establishing the DNA sequencing capacity of the NHGRI Genome Sequencing Program. The cost-accounting data presented by NHGRI is on the basis of (1) "Cost per megabase" (Mb: 1 million base) of DNA sequence (Fig. 2.10a); and (2) "Cost per genome" (human-size genome) (Fig. 2.10b). In each represented graph, a comparison has been made between the cost data and Moore's Law, which describes a long-term trend in the computer hardware industry that involves the doubling of "compute power" every 2 years. Technology improvements that "keep up" with Moore's Law are widely regarded to be doing exceedingly well, making it useful for comparison (National Human Genome Research Institute.genome.gov). The reduction of the cost of DNA sequencing per megabase and per genome from 2001 until 2012 is also represented in Table 2.2.

2.10   Sequencing Status

The number of completed sequenced genomes and the number of genome sequencing projects that are in progress are reported daily by Genome online database (GOLD). Table 2.3 shows the sequencing status until March 29, 2013.

Table 2.1 Comparison of available next-generation sequencing methods' characteristics

Instrument	Read length (nucleotides)	No. of reads[a]	Output (Gb)[a]	No. of samples[a,b]	Runtime	Advantages	Disadvantages
Roche 454 GS FLX+	700[c]	1×10^6	0.7	192[d]	23 h	Long reads, short run time	Homopolymer errors, expensive
Illumina HiSeq2000	100[e]	3×10^9	600	384	11 days[f]	High yield	No. of index tags limiting
Life technologies SOLiD 5500xl	75[g]	15×10^9	180	1,152	14 days[f]	Inherent error correction	Short reads[g]
Roche 454 GS junior	400[c]	1×10^5	0.035	132	9 h	Long reads	Homopolymer errors, expensive
Illumina MiSeq	150	5×10^6	15	96	27 h	Short run time, ease of use	Expensive per base
Ion torrent PGM ion 316 chip	>100[h]	1×10^6	0.1	16	2 h	Short run time, low reagent cost	Not well evaluated
Helicos biosciences HeliScope	35[h]	1×10^9	35	4,800	8 days	SMS, sequences RNA	Short reads, high error rate
Pacific biosciences PacBio RS	>1,000[h]	1×10^5	0.1	1	90 min	SMS, long reads, short run time	High error rate, low yield

[a] Numbers calculated for two flow cells on HiSeq2000 and SOLiD 5500xl

[b] Calculated as no. of index tags (provided by the sequencing company) × no. of divisions on solid support

[c] Average for single-end sequencing, paired-end reads are shorter

[d] No. of reads decreases when the PicoTiterPlate is divided

[e] 36 nucleotides for mate-pair reads

[f] Run time depends on the read length, and on whether one or two flow cells are used

[g] Second read in paired-end sequencing is limited to 35 nucleotides, and mate-pair reads to 60 nucleotides

[h] Average

SMS = single molecule sequencing

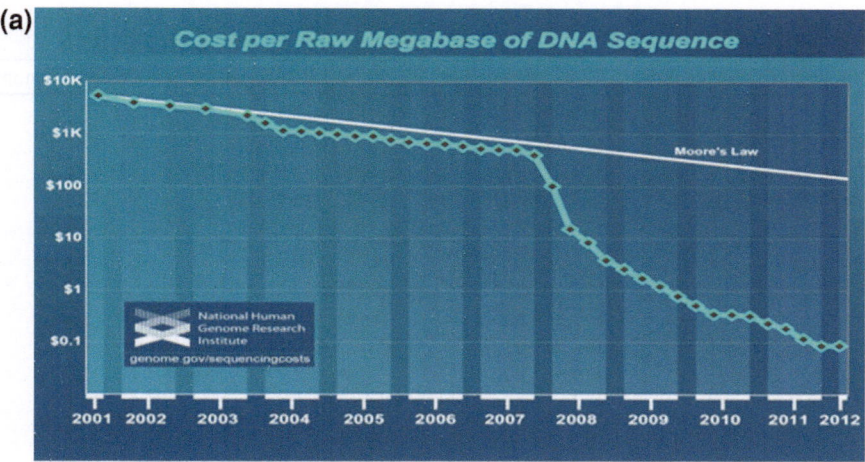

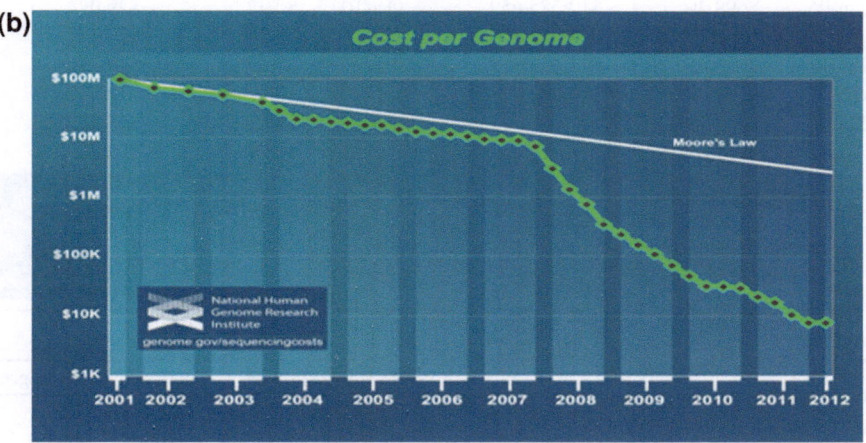

Fig. 2.10 Reduction in the cost of sequencing between 2001 and 2012. **a** The graph of the cost per megabase of DNA and **b** The graph of the cost per human-size genome. Also in each graph is hypothetical data reflecting Moore's Law (National Human Genome Research Institute.genome. gov)

2.11 Shortcoming of NGS Techniques: Short-Reads and Reads Accuracy Issues

The common problem with all current NGS technologies is the short length of reads (sequenced fragments) and higher error rate than those of the traditional Sanger sequencing method. The short-reads that are produced in these methods are especially problematic when large DNA fragments, e.g., a whole genome, is to be sequenced [1]. This drawback is an issue especially in sequencing new genomes

Table 2.2 Reduction in the cost of sequencing per megabase and per genome between 2001 and 2012

Date	Cost per Mb	Cost per genome	Date	Cost per Mb	Cost per genome
Sep-01	$5,292.39	$95,263,072	Jul-07	$495.96	$8,927,342
Mar-02	$3,898.64	$70,175,437	Oct-07	$397.09	$7,147,571
Sep-02	$3,413.80	$61,448,422	Jan-08	$102.13	$3,063,820
Mar-03	$2,986.20	$53,751,684	Apr-08	$15.03	$1,352,982
Oct-03	$2,230.98	$40,157,554	Jul-08	$8.36	$752,080
Jan-04	$1,598.91	$28,780,376	Oct-08	$3.81	$342,502
Apr-04	$1,135.70	$20,442,576	Jan-09	$2.59	$232,735
Jul-04	$1,107.46	$19,934,346	Apr-09	$1.72	$154,714
Oct-04	$1,028.85	$18,519,312	Jul-09	$1.20	$108,065
Jan-05	$974.16	$17,534,970	Oct-09	$0.78	$70,333
Apr-05	$897.76	$16,159,699	Jan-10	$0.52	$46,774
Jul-05	$898.90	$16,180,224	Apr-10	$0.35	$31,512
Oct-05	$766.73	$13,801,124	Jul-10	$0.35	$31,125
Jan-06	$699.20	$12,585,659	Oct-10	$0.32	$29,092
Apr-06	651.81	$11,732,535	Jan-11	$0.23	$20,963
Jul-06	$636.41	$11,455,315	Apr-11	$0.19	$16,712
Oct-06	$581.92	$10,474,556	Jul-11	$0.12	$10,497
Jan-07	$522.71	$9,408,739	Oct-11	$0.09	$7,743
Apr-07	$502.61	$9,047,003	Jan-12	$0.09	$7,666

National Human Genome Research Institute
genome.gov/sequencingcosts

Table 2.3 Reported number of completed and in-progress genome sequencing projects

	Complete	In progress
Archea	268	70
Bacteria	6,998	80,771
Eukarya	1,230	1,399

(with no prior knowledge of the relative position of fragments) and in sequencing highly rearranged genome segments such as one might discover in cancer genomes or in regions of structural variation. However, as mentioned before, using a paired-end sequencing approach would be beneficial to the utility of short-reads for sequencing de novo sequence assembly and for sequencing rearranged genomic segments [21]. Besides the paired-end sequencing technique, several new assembly algorithms began to be designed recently in order to overcome this problem [13]. The quality of each base is also lower than that of the Sanger chemistry reads. This problem was solved for the Sanger sequencing method through the validated base quality score originating from Phred programs. The quality scores compress a variety of types of information about the quality of base calls into a readily usable probability-of-error value. To deliver accurate results, many analysis tools and all assemblers require quality score input. To date, the quality scores offered by new sequencing technology companies are not good substitutes for Phred scores. The accurate quality score is a key issue in various areas, such as sensitive and specific polymorphism detection, enabling accurate statistical modeling of the significance of read alignments and providing a quantitative basis for comparison of sequencing results from different technologies [22].

The template preparation process in all NGS methods involves "shearing the genomic DNA into small fragments" or "creating paired-end libraries." The common feature that is observed in all NGS methods is that the resulting DNA templates are bound to the solid surface with specific distances from each other that allow millions of sequencing reaction to be performed in parallel [1].

2.12 NGS File Formats

Besides a few archives available for next-gen sequence data like NCBI's Sequence Read Archive or SRA (http://www.ncbi.nlm.nih.gov/Traces/sra/sra.cgi), ENA-Reads at EBI (http://www.ebi.ac.uk/embl/Documentation/ENA-Reads.html), and the DNA Data Bank of Japan (http://www.ddbj.nig.ac.jp/sub/trace_sra-e.html), there is other next-gen sequence data created and hosted by many authors and researchers [23].

Common next-generation sequencing file types among different data analysis and visualization tools are listed in Table 2.4 [23].

FASTQ format is the most common format for raw reads, as is BAM for aligned reads. FASTQ files are only kept before the alignment is done and the BAM file is generated.

2.13 NGS Applications

NGS application areas are defined in Table 2.5 (below) in terms of the particular source of nucleic acids selected for sequencing as well as the analysis strategy chosen to interpret the sequence information. A brief description of this application will be presented later. (Please see references for more details.)

RNA-Seq and small RNA sequencing

RNA-Seq, which was introduced as a new way for gene expression analysis, refers to sequencing the complex mixture of transcripts that were initially converted to cDNA using reverse transcription. RNA-Seq has resulted in the production of new data; for instance, finding novel spliced junctions, antisense regulation mode of action, or intragenic expression are among the many applications of RNA-Seq [24]. The term "small RNA" refers to microRNA and other non-coding RNAs. Sequencing of small RNA allows for accurate detection and quantification of rare small RNA sequences. Differential expression of known microRNAs, as well as finding novel microRNA targets, is also possible using NGS methods.

Table 2.4 NGS data formats

File extension	Description	Reference	Publication
.fasta	Classic DNA sequence file format	http://www.ncbi.nlm.nih.gov/blast/fasta.shtml	n/a
.ace	File format for whole-genome assemblies	Annotated in the documentation for CONSED, currently: http://www.phrap.Org/consed/distributions/README.19.0.txt	Gordon et al. (1998)
.wig	A reference-genome indexed data series for "dense" and continuous data (such as %GC)	http://genome.ucsc.edu/goldenPath/help/wiqqle.html	Kent & Haussler (2002)
.bed	A reference-genome indexed data series for "sparse" data (such as transcriptome data)	http://genome.ucsc.edu/goldenPath/help/bedgraph.html,	Kent & Haussler (2002)
.tab	Tab-delimited text	N/A	n/a
.pdf	Portable document format	Either ISO-32000-1 or http://www.adobe.com/devnet/pdf/pdf_reference.html	n/a
.sam	"Sequence Alignment/Map" format	http://samtools.sourceforge.net/SAM1.pdf	Li (2009)
.bam	Binary format of sam	http://samtools.sourceforge.net/SAM1.pdf	Li (2009)
.fastq	Combination of sequences and quality scores in one file; mainly for data from Illumina sequencers in which case quality scores have been transformed.	http://maq.sourceforge.net/fastg.shtml	Li (2008) and Cock (2010)
.csfasta	Life technologies SOLiD colorspace fasta file—containing color calls (0, 1, 2, 3) rather than base calls	See: http://solidsoftwaretools.com/	n/a
.qual	Per-base quality scores generated during basecalling. All but Illumina scores are scaled to estimate the probability of an incorrect base call, as is in common use for conventional sequencing as Phred quality scores.		n/a
.gff	A flexible format for annotating features (e.g. genes) on a sequence.	http://www.sanger.ac.uk/resources/software/gff/.	n/a
.srf	"Short Read Format"—a new format proposed for short-read DNA sequence	http://srf.sf.net	n/a
.sff	Standard flowgram Format (specific for Roche/454)	http://www.ncbi.nlm.nih.gov/Traces/trace.cgi?cmd=show&f=for mat58tm=docBts=format5#header-global	n/a
.gtf	Gene transfer format, an alternate format to GFF for specifying gene features	http://mblab.wustl.edu/GTF22.html	n/a

Table 2.5 A review of NGS applications

Name	Nucleic acid population	Brief analysis strategy
RNA-Seq	RNA (may be poly-A mRNA or total RNA)	Alignment of reads to "genes"; variations for detecting splice junctions and quantifying abundance
Small RNA sequencing	Small RNA (often miRNA)	Alignment of reads to small RNA references (e.g., miRbase), then to the genome; quantify abundance
ChIP-Seq	DNA bound to protein, captured via antibody (ChIP = Chromatin ImmunoPrecipitation)	Align reads to reference genome, identify peaks & motifs
RIP-Seq	RNA bound to protein, captured via antibody (RIP = RNA ImmunoPrecipitation)	Align reads to reference genome and/or "genes," identify peaks and motifs
Methylation analysis	Select methylated genomic DNA regions, or convert methylated nucleotides to alternate forms	Align reads to reference and either identify peaks or regions of methylation
SNP calling/discovery	All or some genomic DNA or RNA	Either align reads to reference and identify statistically significant SNPs, or compare multiple samples to each other to identify SNPs
Structural variation analysis	Genomic DNA, with two reads (mate-pair reads) per DNA template	Align mate-pairs to reference sequence and interpret structural variants
de novo sequencing	Genomic DNA (possibly with external data e.g., cDNA, genomes of closely related species, etc.)	Piece-together reads to assemble contigs, scaffolds, and (ideally) whole-genome sequence
Metagenomics	Entire RNA or DNA from a (usually microbial) community	Phylogenetic analysis of sequences

ChiP-Seq technique and RIP-Seq

In ChIp-Seq (Chromatin immunoprecipitation), which is used to direct mapping of DNA binding site to a reference genome, Microarray has been replaced by NGS techniques [5]. Protein-bound DNA is captured through interaction of an antibody targeting protein. DNA is then treated with DNase to digest parts of DNA that are not bound to protein. In the next step, protein-bound DNA is separated from protein-part, sequenced and matched to the reference genome. RNA binding protein immunoprecipitation sequencing (RIP-Seq) is an analogue to ChiP-Seq, which is used to identify RNA molecules that are bound to a nuclear or cytoplasmic protein.

Methylation analysis

DNA methylation, which is the major form of epigenetic modification, influences cytosines in promoter sequences that are enriched in CGs (hence referred to as CpG islands) and converts it to 5 methyl cytosine. This modification is known to influence gene expression and it has been understood that specific patterns of methylation are implicated in cancer development. Decoding the genome wide methylation profile plays an important role in finding the correlation between DNA methylation and other epigenetic modifications, and NGS methods have helped in deciphering the methylation profile of the genome [25].

SNP calling/discovery

The discovery of SNPs, including single nucleotide mutation, insertion, and deletion in NGS data, is straightforward, using various methods and software tools. SNP discovery has helped population genetic studies due to the generation of relatively inexpensive, high-depth sequencing data through NGS methods.

Structural variation analysis

SNPs have long been thought to be the most common class of genetic variations and have been widely used in linkage and genome-wide association studies. However, structural variation analysis, which is typically concerned with identification of large-scale amplifications, deletions, translocations, or inversions, is now believed to be widespread in the human genome. NGS has been widely used in structural variation analysis [26].

Metagenomics

One of the early applications of NGS was to determine how all organisms live together in a certain ecosystem, such as in a deep mine or in the ocean. This field is called Meta-genomics; in it, all biological communities of an ecosystem are sequenced en masse to obtain information about all the organisms of that ecosystem. This approach enables the study of communities that could not be easily cultured and thereby provides an unbiased view of the composition and state of the community [27].

2.14 Summary

Recent high-throughput sequencing technologies have their origins in the results of the studies of Sanger's and Gilbert's groups, which introduced two distinct procedures for identifying the order of nucleotides in a DNA fragment in the late 1970s. Among these methods, Sanger's, which is also known as "chain termination" or the "dideoxy" method, became commercialized and has been in use for over 30 years. The massive sequencing projects, such as the Human Genome Project, were also based on Sanger's chemistry. The maximum length of the DNA fragment that can be sequenced by Sanger sequencing technology with high fidelity (1–2 % error rate) is about 1000 bp and as a result of using this technique, shearing the larger pieces of DNA is required. Assembling the sequenced reads is faced with difficulties originated from missing fragments, repeats and intrinsic sequencing errors. Moreover, the cost of the Sanger method is now about $0.5 per 1 kb, which is still too high to sequence a whole human genome for each person ($10,000–$100,000) that can revolutionize the future medicine. The NIH goal that was announced in 2004 was to develop new sequencing technologies that will be able to sequence individual human genomes at a lower cost ($1,000) and, of course, in just a few days.

Since then, many research groups have been working on developing new sequencing systems. First Roche/454 FLX introduced its platform, which was based on a pyrosequencing procedure that was commercially available in 2004. In early 2007, the Illumina (Solexa) Genome Analyzer released its platform, which became highly popular and was based on the bridge PCR. That same year, Applied Biosystems introduced the SOLiD sequencing system, which, in contrast to the two previously mentioned methods, basically works through ligation. All three of these techniques rely on an amplification step before sequencing, which is a costly and time-consuming steps. Ion semiconductor is another NGS method which works on the basis of pH fluctuation. In polonator sequencing technology, sequencing is performed using degenerate nonamers. In 2008, the Helicose sequencing platform, which requires no amplification step and therefore a single DNA molecule that can directly be sequenced became available. Helicose became a prototype for other groups to design new sequencing systems without any amplification step. Nanopore and single molecule real time (SMRT) are the two most promising platforms that are becoming commercialized. With these new sequencing technologies, which have revolutionized the field of genomics, it seems that the $1,000 genome goal will be feasible in the near future. These high-throughput sequencing methods have introduced new fields in biology, which include personal genomics, analysis of RNA transcript, ChIP-seq (Chromatin Immunoprecipitation sequencing), ChIP-chip (Chromatin Immunoprecipitation coupled to DNA microarray) and many others.

References

1. Metzker, M. L. (2009). Sequencing technologies—the next generation. *Nature Reviews Genetics, 11*(1), 31–46.
2. Novák, P., Neumann, P., & Macas, J. (2010). Graph-based clustering and characterization of repetitive sequences in next-generation sequencing data. *BMC Bioinformatics, 11*(1), 378.
3. Dong, H., & Wang, S. (2012). Exploring the cancer genome in the era of next-generation sequencing. *Frontiers of medicine, 6*(1), 48–55.
4. Shendure, J., et al. (2004). Advanced sequencing technologies: Methods and goals. *Nature Reviews Genetics, 5*(5), 335–344.
5. Ansorge, W. J. (2009). Next-generation DNA sequencing techniques. *New Biotechnology, 25*(4), 195–203.
6. Shendure, J., et al. (2005). Accurate multiplex polony sequencing of an evolved bacterial genome. *Science, 309*(5741), 1728–1732.
7. Mardis, E. R. (2008). Next-generation DNA sequencing methods. *Annual Review of Genomics and Human Genetics, 9*, 387–402.
8. Shendure, J., & Ji, H. (2008). Next-generation DNA sequencing. *Nature Biotechnology, 26*(10), 1135–1145.
9. Berglund, E. C., Kiialainen, A., & Syvänen, A. C. (2011). Next-generation sequencing technologies and applications for human genetic history and forensics. *Investigative Genetics, 2*(1), 1–15.
10. Pareek, C. S., Smoczynski, R., & Tretyn, A. (2011). Sequencing technologies and genome sequencing. *Journal of applied genetics, 52*(4), 413–435.
11. Novais, R., & Thorstenson, Y. (2011). The evolution of Pyrosequencing® for microbiology: From genes to genomes. *Journal of Microbiological Methods, 86*(1), 1–7.
12. Kircher, M., & Kelso, J. (2010). High-throughput DNA sequencing–concepts and limitations. *BioEssays, 32*(6), 524–536.
13. Nowrousian, M. (2010). Next-generation sequencing techniques for eukaryotic microorganisms: Sequencing-based solutions to biological problems. *Eukaryotic Cell, 9*(9), 1300–1310.
14. Liu, L., et al. (2012). Comparison of Next-Generation Sequencing Systems. *Journal of Biomedicine and Biotechnology*, 2012.
15. Hui, P. (2012). Next generation sequencing: chemistry, technology and applications. [Without Title], pp. 1–18.
16. Wash, S., & Image, C. (2008). DNA sequencing: generation next–next. *Nature Methods, 5*(3), 267.
17. McNally, B., et al. (2010). Optical recognition of converted DNA nucleotides for single-molecule DNA sequencing using nanopore arrays. *Nano Letters, 10*(6), 2237–2244.
18. Clarke, J., et al. (2009). Continuous base identification for single-molecule nanopore DNA sequencing. *Nature Nanotechnology, 4*(4), 265–270.
19. Tyagi, S., et al. (2000). Molecular beacons: hybridization probes for detection of nucleic acids in homogeneous solutions. Nonradioactive Analysis of Biomolecules, 2nd ed. C. Kessler, ed. Springer-Verlag, Berlin, pp. 606–616.
20. Eid, J., et al. (2009). Real-time DNA sequencing from single polymerase molecules. *Science, 323*(5910), 133–138.
21. Morozova, O., & Marra, M. A. (2008). Applications of next-generation sequencing technologies in functional genomics. *Genomics, 92*(5), 255–264.
22. Brockman, W., et al. (2008). Quality scores and SNP detection in sequencing-by-synthesis systems. *Genome Research, 18*(5), 763–770.
23. Hunicke-Smith, S., Next-generation sequencing data and workflows.
24. Tarazona, S., et al. (2011). Differential expression in RNA-seq: A matter of depth. *Genome Research, 21*(12), 2213–2223.

25. Zhang, Y., & Jeltsch, A. (2010). The application of next generation sequencing in DNA methylation analysis. *Genes, 1*(1), 85–101.
26. Xi, R., Kim, T.-M., & Park, P. J. (2010). Detecting structural variations in the human genome using next generation sequencing. *Briefings in Functional Genomics, 9*(5–6), 405–415.
27. Hugenholtz, P., & Tyson, G. W. (2008). Microbiology: Metagenomics. *Nature, 455*(7212), 481–483.
28. Gordon, D., et al. (1998). Consed: A graphical tool for sequence finishing. *Genome research, 8*(3), 195–202.
29. Kent, W. J., Sugnet, C. W., Furey, T. S., Roskin, K. M., Pringle, T. H., Zahler, A. M., Haussler, D. (2002). *Genome Research, 12*(6), 996–1006.
30. Li, H., et al. (2009). The sequence alignment/map format and SAMtools. *Bioinformatics, 25*(16), 2078–2079.
31. Li, R., et al. (2008). SOAP: Short oligonucleotide alignment program. *Bioinformatics, 24*(5), 713–714.
32. Cock, P. J., et al. (2010). The Sanger FASTQ file format for sequences with quality scores, and the Solexa/Illumina FASTQ variants. *Nucleic acids research, 38*(6), 1767–1771.

Chapter 3
The Assembly of Sequencing Data

Genome science has progressed greatly in recent years and its potential applications caused scientists to believe that biology will be the foremost science of the twenty-first century. The outcome of genome research projects has a major impact on the life sciences. Being able to gain genome sequences can be helpful for other analyses, such as the detection of single nucleotide polymorphisms (SNPs) and comparative genomic research. There are many potential applications for genome research, including molecular medicine, risk assessment, bioarchaeology, anthropology, evolution and human migration, DNA forensics, and agriculture, livestock breeding, and bioprocessing. These applications will not be successfully processed until we are able to sequence genomes within a reasonable amount of time and cost. Personalized medicine is a promising area that is defined as being based on having individual genomes at hand; it is also a big potential market if we look at sequence assembly from this perspective.

With this vast area of applications, there are still only thousands of bacterial genomes, plus a few dozen higher organisms' genomes that are sequenced, and the remainder are still not sequenced. With next-generation sequencing techniques, we can have a genome sequenced in a short time, but even the most advanced sequencing techniques can sequence about 5000 nucleotides. This is while even the smallest viral genomes are made up of several thousand bases (in simple viral genomes) and more complicated genomes are much more larger—for example, about 10^9 base pairs in mammals. Therefore, computational methods are needed to assemble the small sequences needed to form the initial genome. Assembly algorithms are developed to address this problem.

All this shows that there is great potential for using high-quality sequence assembly methods, while the domain knowledge and existing algorithms are not advanced enough to produce good results in the expected amount of time. In this chapter, we'll look at the definition of assembly problems, assembly issues, methods for de novo assembly, and, finally, the methods for evaluating assembly algorithms.

A. Masoudi-Nejad et al., *Next Generation Sequencing and Sequence Assembly*, 41
SpringerBriefs in Systems Biology, DOI: 10.1007/978-1-4614-7726-6_3,
© The Author(s) 2013

3.1 What is De Novo Genome Sequence Assembly?

De novo sequence assembly refers to a computational method for merging the fragments of DNA sequences—obtained from sequencing methods such as those discussed in Chap. 1 of this book—to form longer DNA sequences, and hopefully to reconstruct the primitive genome. In fact, sequencing is done in a way that sequenced streams have some overlaps and therefore can be merged due to their overlaps in order to form larger fragments and finally—and hopefully—the whole genome, when its aim is whole genome assembly.

Assembly can also be performed in a map-based fashion. If the goal of assembly algorithms is to form a new, previously unknown sequence, it is called a de novo assembly algorithm. If there is an existing backbone sequence, and the assembly algorithm just builds a sequence that is similar, not identical, to the backbone sequence, then the algorithm is a mapping. Figure 3.1 [1] graphically demonstrates the differences between sequencing and assembly methods. Section (d) of this figure shows resequencing, i.e., sequencing a known genome, and its assembly, in which the reference genome is available and therefore the assembly task is reduced to genome alignment. In contrast, in the other sections (a–c), the reference genome is not available and the assembly algorithm has to computationally find overlaps between reads and try to assemble them correctly; this is called de novo assembly.

The mapping of resequenced reads to a reference genome is a computationally easier problem than de novo assembly. Several tools are available for mapping reads to the genome, including tools for the mapping of short-reads. MAQ [2], SOAP [3], and SHRiMP [4] are a few examples that use seeding techniques, in addition to a precomputed hash table, for the fast matching of reads to the main

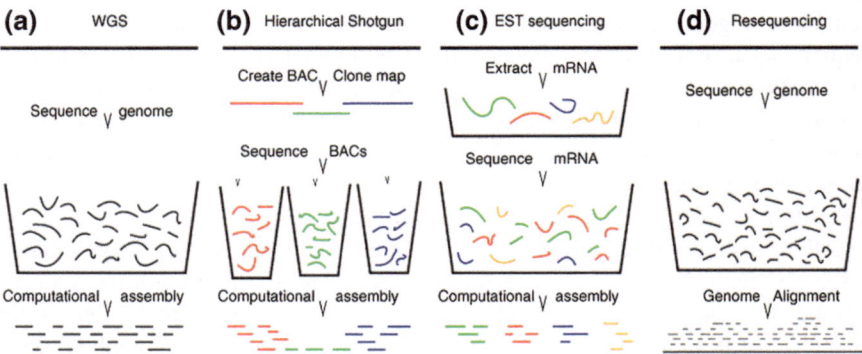

Fig. 3.1 [1] Sequencing methods. Schematic drawing of the four different sequencing procedures. **a** Whole-genome shotgun, where the genome is randomly split into smaller parts and sequenced. **b** Hierarchical shotgun, where a BAC clone map (tilling map) covering the genome is first created, after which the BACs are sequenced. **c** EST sequencing, where mRNA is extracted from tissue and then sequenced. **d** Massively parallel sequencing where short-sequence fragments are aligned to a reference genome

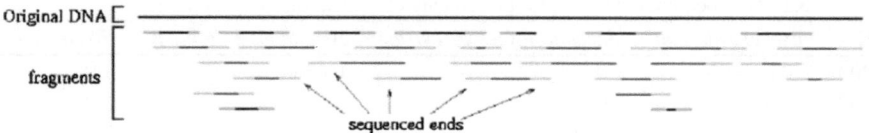

Fig. 3.2 Redundancy in the sequencing phase causes overlaps in the resulting sequence read ends. Each sequence can have overlaps with several other sequences. [http://www.cbcb.umd.edu/research/assembly_primer.shtml]

genome. The mapping is not yet an easy task because of sequencing errors. For example, the ability to detect indels, which occur frequently in 454 sequences, is very limited in the available programs, and most of these tools need subsequent alignment runs in order to be able to detect indels [1]. But still, methods that are capable of using hybrid data inputs, such as paired-end reads to resolve breaks, and also microarray-based genomic selection and multiplex exon capture that helps gain sequences from special locations on the genome sequences, can be nicely used besides mapping algorithms. In this survey, we focus on de novo assembly algorithms.

According to what was described in the Sequencing Methodology Section, sequencing methods can just sequences DNA fragments of length from tens to several thousand bases, depending on the method used. This is while the shortest genomes—related to virus genomes—are at least several thousand base pairs, and for advanced creatures, like plants and animals, it is more than hundreds of millions of base pairs. (The human genome is about 2.9 billion base pairs.) It is easy to understand how difficult the assembly task will be with short-reads gained from next- generation sequencing methods, especially when working with high-throughput next-generation sequencing data, which give in hand, short and erroneous reads.

In order to help the assembly task, the sequencing step is usually done with high coverage. As described in Chap. 1, coverage is the number of reads representing a given nucleotide in the reconstructed sequence. For example, sequencing a DNA with 1,000 base pairs so that 4 fragments of 500-length base pairs are gained, is done with 2X coverage. As can be seen in Fig. 3.2, this redundancy led to sequence overlaps that help the process of assembling DNA fragments. In fact, overlapping regions are the key to finding which fragments should be merged together.

3.2 Challenges of Genome Assembly

It can be understood by comparing the genome size, the sequencing fragment's size, and the large number of reads generated by the sequencing method, that assembly is not a simple task. The first challenge of assembling these sequences is

Table 3.1 The mean number of false placements of k-mers on the genome [5]

K	Escherichia coli	Saccharomyces cerevisiae	Arabidopsis thaliana	Homo sapiens
200	0.063	0.26	0.053	0.18
160	0.068	0.31	0.064	0.49
120	0.074	0.39	0.086	1.7
80	0.082	0.49	0.15	7.2
60	0.088	0.58	0.27	18
50	0.091	0.63	0.39	32
40	0.095	0.69	0.65	78
30	0.11	0.77	1.5	330
20	0.15	1.0	5.7	2,100
10	18	63.8	880	40,000

the existence of too many reads and, therefore, overlaps, which are not necessarily correct overlaps. A small overlap size increases the probability of wrong overlaps. Overlap size is limited to the size of reads as well as the coverage. Table 3.1 [5] shows the mean number of perfect matches of a k-mer in the specified genome, excluding the correct match. This number shows the expected ratio of false overlaps to true ones between reads, which overlap in exactly k bases. The sample size is 10^6 in each case. In fact, this result shows that with smaller reads there is a higher probability of wrongly aligned overlaps [5].

Reads overlapping with more than one other read will form a branch in the resulting assembly. Various ways of choosing between these branches will be further discussed in the algorithm section. One of the methods for dealing with these branches is by using paired-end reads, such as what is done in [6].

There are a few other problems that make assembly even more difficult. One of the major problems is sequencing errors—for example, base pair misreads in the output. Due to sequencing errors, the sequencer's output is in the form of sequences of base calls plus a quality value (QV) for each base call. The use of these quality values can add extra information for the assembly task, but it needs more processing effort and memory; therefore, only some of the assembly algorithms make use of QVs. These low-quality regions of reads are clipped in some of the methods, such as in [7].

Even with preprocessing reads for limiting errors, some errors may still remain in the sequences. Sometimes read errors in read set can be corrected by aligning reads and comparing them (for example, in [8]); however, it is not easy to detect errors from polymorphism variants, and sometimes polymorphism mismatches are also included in computing the error rate in reads [9].

As mentioned in the previous section, in order to assist the assembly task, high coverage data is produced. This coverage is not uniform on the whole sequence and this variability of coverage will lead to a complexity in the assembly task itself. Variability of coverage makes it impossible to use statistical tests and coverage-based analysis to detect repeats [10].

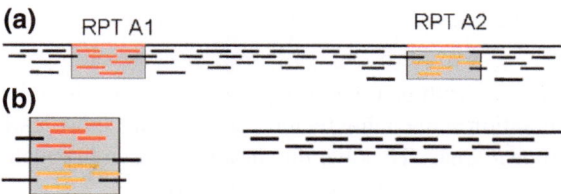

Fig. 3.3 Problem of repeats in genome assembly. **a** If read length is less than repeat length, the reads from repeats are identical to the assembly program. **b** Mis-assembled genome due to a repeat. [http://www.cbcb.umd.edu/research/assembly_primer.shtml]

Another problem in assembling DNA sequences brings us back to the fact that some portions of the genome may remain unsequenced. In this situation—with even perfect assemblers—it is just not possible to handle the situation. As briefly mentioned in Chap. 1, another situation that is very problematic in the assembly of eukaryotic genomes is the presence of (nearly) identical sequences—called "repeats"—in the genome. The existence of repeats makes it difficult to merge the reads in the correct way (Fig. 3.3). If repeat size is less than read size, one can see that there may be many similar reads—from within repeat areas of the genome— that refer to different locations in the main genome.

The problem of repeats can be resolved by high coverage of sequences, but existing errors in sequence data don't allow the repeat discovery task to be very easy. To resolve the repeats that are longer than reads, paired-ends are needed (paired-end [mate-pair] technologies are described in Chap. 1). This is a more complicated task than resolving repeats shorter than read sizes using single reads. Inexact repeats can be separated by the high-stringency alignment of reads and finding read correlations using different base call patterns in them [11]. The task of resolving repeats will be explained later for each assembly algorithm in Chap. 4. All these, in addition to the size of genomes and large number of reads, make assembly a complicated problem requiring an efficient solution and data structure design and computationally high-performance platforms. Intelligent heuristics and tricks play an important role in overcoming these difficulties.

It is important to note that assembling is not applied only to genomic data, and, for example, assembling transcriptomics data (such as expressed sequence tags— "ESTs"), which gives a view of the biological state of a cell, is something that is also very important in practice. However, the challenges in various kinds of data are different. The discontinuity of transcriptomics data results in less contiguity than genomic data. Since repeats mainly exist in intron regions of the genome, repeat is not a major issue in assembling transcriptomics data. But since transcription from a single part of the genome can be done in different patterns (i.e., from different start and end positions), this adds an additional complexity to the assembly of transcriptomics data. Algorithmic approaches are needed to handle other situations referring to ESTs—for example, different rate of expression (highly expressed genes), alternative splicing, and paralogous genes. These problems are even more serious with the contamination of the CDNA library by genomic data [12].

3.3 Use of Paired-End Reads in the Assembly

Sequence assembly algorithms use overlaps between reads in order to merge them again. Next- generation sequencing techniques generate short-reads, which is very problematic in repeat areas. To solve this problem, assembly methods usually use mate-pair or paired-end libraries (as discussed earlier).

Mate-pair (or paired-end reads) are DNA fragments with both ends sequenced. In other words, they can be thought of as two reads with a specific distance between them. As explained, the sequence reads from NGS methods are short. When reads are shorter than the length of repeated areas of the genome, we usually cannot resolve repeat areas in the assembly phase (Fig. 3.3). A mate-pair fragment can be longer than NGS reads, since there is a gap or un-sequenced part between two sequenced parts, and therefore these fragments can be used to resolve repeats. Various protocols are usually used to generate mate-pairs and paired-end reads. As previously mentioned, the difference between the two applications is in their length and generating technology; mate-pairs are about 2–5 KB, while paired-end reads are much shorter, rarely more than 500 bp. There are various methods for constructing paired-end and mate-pair libraries. In (IlluminaMatePair,[1] IlluminaPairedEnd[2]), the construction method of Illumina technology is briefly described. Long-insert paired-end libraries are useful for de novo sequencing and genome finishing (which will be explained later in this section). Combining data generated from the mate-pair library sequencing with that of short-insert paired-end reads provides a powerful combination of read lengths for maximal genomic sequencing coverage across the genome.[3]

3.4 Data Preprocessing Methods and Sequence Read Correction Methods

Usually when data is not clean, i.e., not ready to be used by the main algorithm, a preprocessing step is needed to prepare it. Some assembly algorithms may also include the error handling process in the assembly, which leads to computational and memory overhead [13].

As mentioned in Sect. 1.2, the output of sequencing is not error-free, and each sequencing platform has its own issues and errors in the final read library. There are also other problems with data; for example, the existence of similar reads from within repeats, which are not from the same part of the genome, may mislead the assembly algorithm. Therefore a preprocessing step is defined in the assembly

---

[1] IlluminaMatePair: http://www.illumina.com/technology/mate_pair_sequencing_assay.ilmn

[2] IlluminaPairedEnd: http://www.illumina.com/technology/paired_end_sequencing_assay.ilmn

[3] http://www.illumina.com/technology/mate_pair_sequencing_assay.ilmn

algorithm. Some algorithms use quality score values in order to filter the input data. In this case, it is important to know what variant of quality score data is encoded [14]. It is good to know that it may not be necessarily to do this correction and preprocessing in some algorithms, for example score-based trimming and filtering had not led to a better result in Velvet [12].

Some of the assembly algorithms, like SSAKE [15], are designed for error-free reads and SHARCGS [16] is designed to work on reads with a low error rate (below 0.05 %).Therefore these methods need a preprocessing of the reads prior to assembly in order to filter erroneous reads.

In some algorithms, a trimming is done on sequence reads in order to cut the erroneous parts of each read before they can be used as an input for the assembly algorithm. In some other preprocessing methods, a portion of reads (i.e., the reads themselves), which are likely to lead to a misleading in the assembly algorithm, are removed from the read set. The read screening phase in [6] is performed for this purpose. "Solid reads" (i.e., error-free and non-repetitive reads) are identified using this read screening process. Figure 3.4 shows a statistical analysis of read size in [6]. In this analysis:

- If a k-mer occurs once, it is likely to be a sequencing error.
- If a k-mer occurs too many times, it is likely to be a repeat.

The area between the two determined spots in the diagram shows the interval in which solid reads exist. The solid reads are chosen as starting points for read extensions in this algorithm. In fact, regions with high depth are considered to be repeats and they are so identified in order to be used properly in the assembly

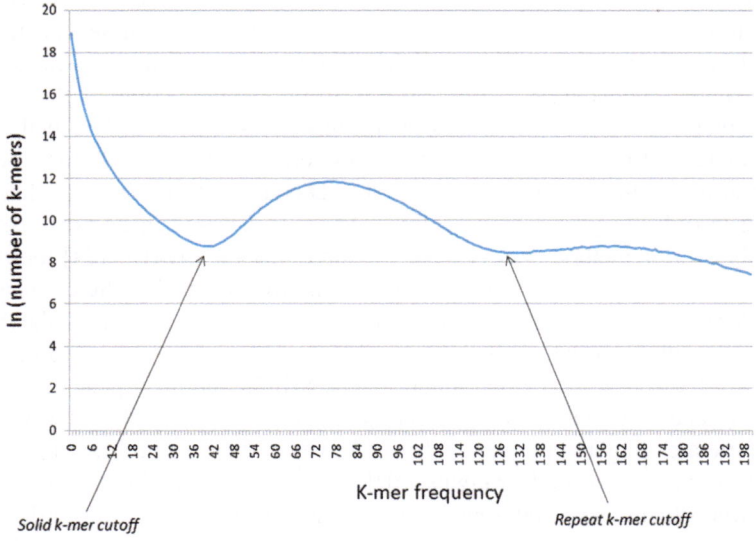

Fig. 3.4 Statistical analysis for filtering erroneous reads [6]

algorithm. The early use of repeated regions will likely mislead the assembly, and it is better to postpone their use in the final steps of assembly. This is in fact the strategy used by PE-ASSEMBLER. Another way to resolve repeated reads is to compare them with known repeats in the genome, in the case where we have a reference genome. There are tools that are specially designed for the task of repeat masking, such as RepeatMasker [17], which uses curated repeat databases to identify and mask the repeats. Besides these filtering methods for curating repeats, there are some methods that try to correct reads of executing the main assembly algorithm beforehand.

Reference [18] uses a preprocessing phase to correct reads prior to assembly. This correction is done to turn many reads into error-free ones. The correction is done using an algorithm that is a modified version of Spectral Alignment described in Pevzner et al. [19] and Chaisson and Pevzner [20]. In this preprocessing method, first a set of all sufficiently frequent reads, called 1-tuples, are selected and their spectrum is computed. Then for each read r, a string r* is computed, which is the sequence with a minimized distance (for example, the Hamming distance) from r, such that all 1-tuples belonging to r* are in the spectrum. SOLiDAccuracy Enhancement Tool (SAET, http://solidsoftwaretools.com/gf/project/saet/) and preprocessing phases in methods like [5] and [21] use the same approach for error-correction. Alignment is also used for error-correction. The alignment approach, which is more suitable for Sanger sequencing reads, uses multiple sequence alignment of the reads in order to find similar reads and detect and correct errors in them. This approach is computationally expensive since it necessitates the aligning of millions of reads. MisEd [22] and the preprocessing step in ARACHNE [8] use the alignment approach for error-correction of reads.

As mentioned above, there are some tools, like SAET, that were specially developed for preprocessing the reads and correcting the sequencing reads. These tools are designed to perform error-correction on the output of sequencing platforms based on the characteristics of the platform. Most of the tools make use of only a single sequencing platform, but it is also possible to combine the output of several platforms in order to better perform this error-correction. The method used by Salmela [23] is an example of sequencing error-correction based on the combining of data sources, which combines SOLiD color space reads with other forms of data in order to gain a high-quality read set to be used in de novo assembly algorithms. This method is developed by improving the SHREC [24] tool, which allows us to utilize both base space (such as the Roche/454 Life Sciences and the SOLEXA/Illumina sequencing platform) and color space (such as the SOLiD sequencing platform, which is based on color coding of the reads) to perform a better error-correction on equal length reads. SHREC builds a generalized suffix trie from reads and then tries to correct them at the trie intermediate levels. Nodes in this trie are weighted by the number of leaves in their own rooted subtrie. By a statistical analysis on node weights, SHREC identifies possible erroneous reads and corrects them using nodes having reads with suffixes similar to subtries rooted at the erroneous node. In [23] the statistical model's applicability is verified for variable length reads. By a combination of base space and color space, it builds a generalized suffix trie and performs error-correction for detecting and possibly correcting insertions, deletions, and substitutions.

As mentioned in Sect. 1.2, ESTs are another type of data that can be the input of an assembly algorithm. When assembling ESTs, a clustering phase is usually performed in order to cluster ESTs from the same gene together. This can lead to an over-clustering in EST data, a condition in which ESTs from various genes are wrongly clustered into one cluster. This happens in the cloning procedure when two originally separate sequences are mistakenly put into the same read; it is called "chimerism." Also, it is possible to cluster paralogous genes together since they are highly similar. These problems can be solved by the use of suitable clustering algorithms, like single transitive single linkage clustering [25], or methods such as double linkage of geneDistiller [26]. Another problem in EST libraries, that of the existence of highly expressed genes, can also be addressed in the preprocessing phase. The removal of known housekeeping genes, adding annotated gene sequences as a template for ESTs, and seeded clustering, are among the solutions to this problem [1].

Even with these strategies, deep clusters (i.e., clusters with a high number of ESTs) may remain in the cluster set and should be treated in a suitable way, similar to what is done in "containment clustering" of TGICL [27] and in geneDistiller [26] by an alignment/consensus strategy [1].

The tools and methods in this section were designed for handling errors in sequencing result. This is a preparing-step for the assembly task. This is another kind of error in the assembly process that we have to be careful about. While errors in this section were about sequencing errors and the input of assembly algorithms, the other type returns to errors existing in assembling sequences, i.e., the output of an assembly algorithm. This kind of error is handled in finishing and genome-completion steps, which is a costly task and is discussed in the next section.

3.5 Assembly Errors

The outputs of assembly algorithms, i.e., assembled genome sequences, are the fundamentals of genome research. But genome sequencing and assembly methods are not yet exact enough to give in hand complete and errorless gap-free genome sequences. Some of the problems return to the incompleteness of the sequence reads to be assembled, and others to the inefficiency of assembly algorithms to find the right assembly correctly. As a result, there are only a few genomes, including some mammals (for example, humans, mice and dogs) that are considered to be complete and error-free, i.e., having approximately one error in 10^4 bases, and no gap [28]. The task of genome completing is a very costly one, even though it is necessary to remove errors and fill the gaps to make these assembled sequences usable in many of areas of genome research. But what are the errors in an assembled genome and how they can affect the usability of genome sequences?

Any mistake in assembly, such as insertion or deletion, substitution of correct bases with incorrect ones, or any translocation or other imperfections, is considered to be an error and can mislead further analysis of this data [29, 30, 31]. In some usages, such as finding protein-coding regions, insertions and deletions can mislead the result completely by transforming codon frames. Even in non-coding regions, sequencing errors may mislead the analyses from finding conserved and functional parts of the genome [32]. This shows the importance of the gap-filling and error-correction phase in assembly. Also, when a genome sequence is finished, it is necessary to have a method for evaluating the quality of the assembly.

Many genome assembly algorithms and tools have been developed to be used in various sequencing platform libraries. Examples of these methods are Celera Assembler [33], PCAP [34], and ARACHNE [8]. By introducing next-generation sequencing methods, new algorithmic issues arose due to the short length and high error rate of the output reads; therefore new tools, like Velvet [35], ALLPATHS [5] or MAQ [2], were developed to deal with this situation. Now the question is how one can evaluate these methods or compare them with each other. Besides differences in the input type of data (sequencing platform, sequencing error handling, etc.), is there any way to compare their quality—or, in other words, to estimate the assembly errors? This is the topic of the next section.

3.6 Evaluation of Assembly Methods

In order to evaluate the output of an assembly algorithm, we need to have some criteria. In [36], two categories of features are used to evaluate assemblies: features based on correctness scores, and features based on size statistics. In fact, these can be seen as main categories used for evaluating assemblies. Correctness scores are features that show how well an assembly is matched to the correct genome. In other words, this group shows the accuracy of the final assembly; this kind of features needs a reference genome. The other criteria—size statistics—are defined based on the high contiguity goal of assemblers. One of the main metrics in this category, which is used to evaluate many assembly methods, is N50, which is the number of longest contigs, whose length exceeds 50 % of the total length of contigs. This means that if we order contigs by their length and select the first n contigs until their size goes beyond 50 % of the total length, that n will be equal to N50. Another form of N50 that is more commonly used is N50 contig size, which is the scaffold or contig length, such that 50 % of the assembled sequences lie in scaffolds of this size or larger [37]. N50 can quantify the assembler's ability to make large contigs and is not able to capture other aspects of its quality.

In fact, there is a trade-off between accuracy and contiguity measures [38]. Maximizing one of them will lead to minimizing the other. For example, lowering the thresholds for allowing more small-sequence blocks to merge will result in more incorrectly assembled parts [32]. The Feature-response curve (FRC), a metric introduced by Narzissi and Mishara [39], captures the quality and contig size trade-off. This is similar to receiver-operating characteristic (ROC) curves for comparing the performance of statistical inference methods. FRC demonstrates how well an assembler is able to exploit the relationship between incorrectly assembled contigs (false positives, contributing with features) against gaps in the assembly (false positives, contributing to a fraction of genome-coverage or "response"), with all parameters (like read-length, sequencing error, or depth) kept constant [39]. FRC does not need any reference sequence for evaluation, and therefore can be used for de novo assembly algorithms, but it has still some shortcomings, like considering some types of assembly errors with uniform weighting.

N50 is a global measure for attaining assembly quality, while it fails to give information about fine scale accuracy measures—for example, if bases are deleted, inserted or substituted incorrectly in the assembly. A problem with accuracy measurement at this level is that it needs the main genome in order to have the assembly compared with it (for example, by alignment and comparison of assembled and reference genomes), and there is still a need for an accurate measure that can be determined independently of the whole-genome itself. Some methods, like [32], use statistical analysis of closely related species in order to quantify error rates in assembled genomes. This comparative genomics approach makes the validation process independent of the existence of a genome itself, and uses closely related species in alignment steps. In such a method, there is a need for statistical analysis in order to be able to separate incorrect indel errors from true evolutionary ones. Finally, another factor that may be taken into consideration in comparing assembly methods is the computational time and memory used. But these are not criteria for evaluating the assembly itself, and we can ignore computational time in favor of gaining more accurate assemblies. There are some recommendations in the literature like that which is given in [12], which can be taken into consideration when choosing an assembly algorithm.

3.7   Summary

The process of sequencing and assembly can be summarized as shown in Fig. 3.5. In the first step, after the genome sequence is selected and fragmented (and possibly amplified, based on the sequencing method), it is given to a sequencer. Sequencers will produce raw data for assembly. This raw data is in a special format, for example, in color space or base space. Base calling is the process that generates reads from this base, which is called raw data. Each base has a quality value in this step, which rises up due to sequencing errors. Other techniques may be used to correct or filter some reads in the raw data. This error detection and curation were explained in Sect. 3.4, in data preprocessing, and sequence read correction methods. The sequence reads, which may be in the form of simple or paired-end reads, may be used in direct mapping when the genome sequence is known to us. If the genome sequence is not previously known, de novo assembly

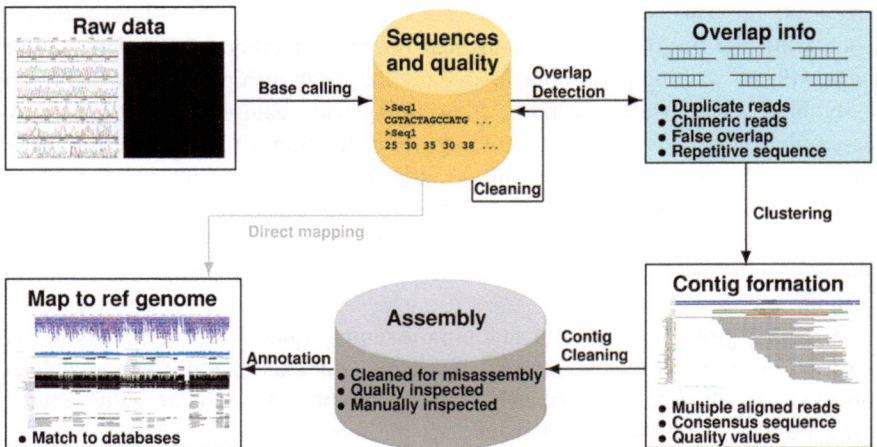

Fig. 3.5 [1] Assembly pipeline. The typical pipeline of a sequencing project. Sequenced reads are generated, after which they are cleaned and assembled. Following the assembly, annotation and analysis can be performed. The *gray line* shows the pipeline for massively parallel sequencing where the reads are mapped to a reference genome, while the full pipeline is for de novo sequencing and assembly [1]

algorithms will be used instead for assembling the reads. This process includes overlap finding (which is useful in greedy algorithms or for forming a graph from reads), contig formation, and then the final step of assembly, which may be guided manually in order to close gaps and finish the assembly. Following the assembly, annotations may be done with the aid of existing databases.

In the next section of this book, we'll have a review of the main approaches used in assembly algorithms, and then we'll explain a number of them in detail in order to better clarify how each approach works.

References

1. Scheibye-Alsing, K., et al. (2009). Sequence assembly. *Computational Biology and Chemistry, 33*(2), 121–136.
2. Li, H., Ruan, J., & Durbin, R. (2008). Mapping short DNA sequencing reads and calling variants using mapping quality scores. *Genome Research, 18*(11), 1851–1858.
3. Li, R., et al. (2008). SOAP: Short oligonucleotide alignment program. *Bioinformatics, 24*(5), 713–714.
4. Rumble, S. M., et al. (2009). SHRiMP: Accurate mapping of short color-space reads. *PLoS Computational Biology, 5*(5), e1000386.
5. Butler, J., et al. (2008). ALLPATHS: De novo assembly of whole-genome shotgun microreads. *Genome Research, 18*(5), 810–820.
6. Ariyaratne, P. N., & Sung, W. K. (2011). PE-Assembler: De novo assembler using short paired-end reads. *Bioinformatics, 27*(2), 167–174.

7. Huang, X., & Madan, A. (1999). CAP3: A DNA sequence assembly program. *Genome Research, 9*(9), 868–877.
8. Batzoglou, S., et al. (2002). ARACHNE: A whole-genome shotgun assembler. *Genome Research, 12*(1), 177–189.
9. Simpson, J. T., et al. (2009). ABySS: A parallel assembler for short read sequence data. *Genome Research, 19*(6), 1117–1123.
10. Miller, J. R., Koren, S., & Sutton, G. (2010). Assembly algorithms for next-generation sequencing data. *Genomics, 95*(6), 315.
11. Kececioglu, J., & Ju, J. (2001). Separating repeats in DNA sequence assembly. *Proceedings of the Fifth Annual International Conference on Computational Biology*, ACM.
12. Paszkiewicz, K., & Studholme, D. J. (2010). De novo assembly of short sequence reads. *Briefings in Bioinformatics, 11*(5), 457–472.
13. Shi, H., et al. (2010). Quality-score guided error correction for short-read sequencing data using CUDA. *Procedia Computer Science, 1*(1), 1129–1138.
14. Cock, P. J. A., et al. (2010). The sanger FASTQ file format for sequences with quality scores, and the Solexa/Illumina FASTQ variants. *Nucleic Acids Research, 38*(6), 1767–1771.
15. Warren, R. L., et al. (2007). Assembling millions of short DNA sequences using SSAKE. *Bioinformatics, 23*(4), 500–501.
16. Dohm, J. C., et al. (2007). SHARCGS, a fast and highly accurate short-read assembly algorithm for de novo genomic sequencing. *Genome Research, 17*(11), 1697–1706.
17. Smit, A., Hubley, R., & Green, P. (2004). *RepeatMasker Open-3.0* 1996–2004. Institute for Systems Biology, Seattle.
18. Chaisson, M. J., Brinza, D., & Pevzner, P. A. (2009). De novo fragment assembly with short mate-paired reads: Does the read length matter? *Genome Research, 19*(2), 336–346.
19. Pevzner, P. A., Tang, H., & Waterman, M. S. (2001). An Eulerian path approach to DNA fragment assembly. *Proceedings of the National Academy of Sciences, 98*(17), 9748–9753.
20. Chaisson, M. J., & Pevzner, P. A. (2008). Short read fragment assembly of bacterial genomes. *Genome Research, 18*(2), 324–330.
21. Li, R., et al. (2010). De novo assembly of human genomes with massively parallel short read sequencing. *Genome Research, 20*(2), 265–272.
22. Tammi, M. T., et al. (2003). Correcting errors in shotgun sequences. *Nucleic Acids Research, 31*(15), 4663–4672.
23. Salmela, L. (2010). Correction of sequencing errors in a mixed set of reads. *Bioinformatics, 26*(10), 1284–1290.
24. Schröder, J., et al. (2009). SHREC: A short-read error correction method. *Bioinformatics, 25*(17), 2157–2163.
25. Quackenbush, J., et al. (2000). The TIGR gene indices: Reconstruction and representation of expressed gene sequences. *Nucleic Acids Research, 28*(1), 141–145.
26. Gilchrist, M. J., et al. (2004) Defining a large set of full-length clones from a < i > Xenopus tropicalis </i > EST project. *Developmental Biology, 271*(2), 498–516.
27. Pertea, G., et al. (2003). TIGR Gene Indices clustering tools (TGICL): A software system for fast clustering of large EST datasets. *Bioinformatics, 19*(5), 651–652.
28. Church, D. M., et al. (2009). Lineage-specific biology revealed by a finished genome assembly of the mouse. *PLoS Biology, 7*(5), e1000112.
29. Salzberg, S. L., & Yorke, J. A. (2005). Beware of mis-assembled genomes. *Bioinformatics, 21*(24), 4320–4321.
30. Choi, J. H., et al. (2008). A machine-learning approach to combined evidence validation of genome assemblies. *Bioinformatics, 24*(6), 744–750.
31. Phillippy, A. M., Schatz, M. C., & Pop, M. (2008). Genome assembly forensics: Finding the elusive mis-assembly. *Genome Biology, 9*(3), R55.
32. Meader, S., et al. (2010). Genome assembly quality: Assessment and improvement using the neutral indel model. *Genome Research, 20*(5), 675–684.
33. Myers, E. W., et al. (2000). A whole-genome assembly of Drosophila. *Science, 287*(5461), 2196–2204.

34. Huang, X., et al. (2003). PCAP: A whole-genome assembly program. *Genome Research,* *13*(9), 2164–2170.
35. Zerbino, D. R., & Birney, E. (2008). Velvet: Algorithms for de novo short read assembly using de Bruijn graphs. *Genome Research, 18*(5), 821–829.
36. Haiminen, N., et al. (2011). Evaluation of methods for de novo genome assembly from high-throughput sequencing reads reveals dependencies that affect the quality of the results. *PLoS One, 6*(9), e24182.
37. Namiki, T., et al. (2011). MetaVelvet: An extension of velvet assembler to de novo metagenome assembly from short sequence reads. *Proceedings of the 2nd ACM Conference on Bioinformatics, Computational Biology and Biomedicine*, ACM.
38. Ewing, B., & Green, P. (1998). Base-calling of automated sequencer traces using Phred. II. error probabilities. *Genome Research, 8*(3), 186–194.
39. Narzisi, G., & Mishra, B. (2011). Comparing de novo genome assembly: The long and short of it. *PLoS One, 6*(4), e19175.

Chapter 4
De Novo Assembly Algorithms

In an ideal case, an assembly algorithm should merge overlapped reads to one long continuous sequence, called contig, which is a chromosome in the primitive genome. But due to sequencing errors and the existence of unsequenced parts, contigs gained from the assembly algorithm are not complete enough to form chromosomes. Even with high coverage, there is still a non-zero probability for the existence of unsequenced parts and sequencing errors. The ability of the assembler to form contigs is also affected by repeated regions in the genome. As shown in Fig. 3.3 in the previous chapter, two parts of different repeat areas are mapped to one in the assembler because of the weakness of repeat detection in the assembler. Figure 4.1 shows how a typical assembly algorithm works in overlap detection and contig generation phases.

As previously explained, paired-end read libraries are other available data for assemblers. This data can be useful to extend contigs and also resolve repeat areas. The task of ordering and orienting contigs along a chromosome using paired-reads is called scaffolding (Fig. 4.2). If one end of a paired-read is assembled in a contig and the other end in a second contig, it can be inferred that these contigs are adjacent in the final assembly.

It is important to note that paired-read data is not accurate and may contain inconsistencies. In the initial genome assembly project for *Drosophila melanogaster* [1], 10–20 % of paired-read information was said to be false. For this reason, methods use a way in order to increase the confidence of used information, like that which is done in SOAPdenove [2], which accepts a defined order and distance between contigs if there are at least three paired-reads proofing them, or ABySS [3], which uses at least five read-pairs for this.

Most assembly algorithms contain a scaffolding phase, but there are also stand-alone scaffolding tools such as Bambus [4].

Scaffolding can also make use of whole genome mapping data. Optical mapping [5] is a good example of this. An approximation of locations of restriction enzyme cuts along the genome can be determined by optical mapping. This information can be used to find the location of contigs in the genome. SOMA [6] is a program that is

A. Masoudi-Nejad et al., *Next Generation Sequencing and Sequence Assembly*,
SpringerBriefs in Systems Biology, DOI: 10.1007/978-1-4614-7726-6_4,
© The Author(s) 2013

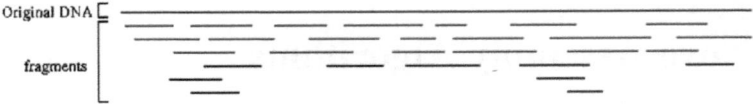

(a) Original DNA broken into a collection of fragments

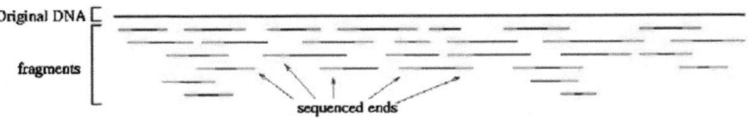

(b) The end of each fragment (drawn in green) are sequenced

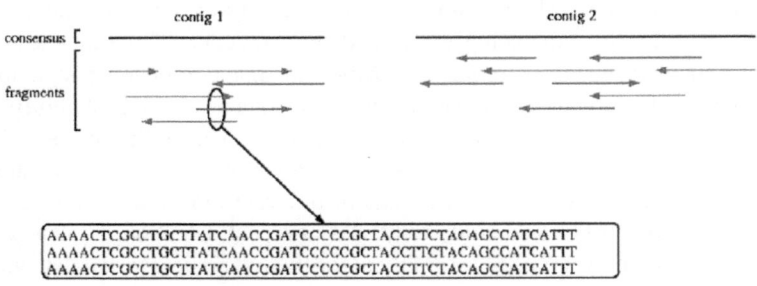

(c) Merging reads to form contigs.

Fig. 4.1 Typical steps in an assembly process. **a** Breaking of original DNA into small sequences that are suitable to be read by sequencers. **b** Sequencing ends of fragments depending on the goal of sequencing (i.e., we want a paired-end library or simple reads). **c** Merging reads to form contigs. [http://www.cbcb.umd.edu/research/assembly_primer.shtml]

Fig. 4.2 A scaffold of 3 contigs (the *thick arrows*) held together by mate-pairs. *Thin lines* connect the paired-ends. [http://www.cbcb.umd.edu/research/assembly_primer.shtml]

designed to generate an in silico restriction map of contigs and use the mapping of generated maps against genome-wide optical maps for scaffolding.

Even after scaffolding, there are rare conditions that lead to chromosome sets of the main genome. There are still gaps in the assembled data, which is the target of the next step of the assembling task, called the "finishing" (or "gap closure" or "gap-filling") step. In this step, additional laboratory experiments and manual curation are performed to validate the correctness of the final assembly, which leads to a more exact reconstruction of the original genome.

Many of the assembly algorithms use pre-processing methods for improving the accuracy of the algorithm. Methods for correcting reads or discarding ambiguous reads, and filtering reads with low coverage or low quality, are some examples. In the next section we'll see some of these examples in existing assembly algorithms.

4.1 Mapping Assembly to a Graph Problem

A graph is an abstract representation of a set of objects, called "vertices" or "nodes" of a graph. There can be connection between every two objects in a graph. A link between two graph nodes is called an edge; it can be directed (so they define an order between connecting nodes), or undirected. The edges represent the connection between relative nodes.

In modelling the assembly as a graph, reads are usually modelled as graph nodes, and edges connect nodes with overlap in the graph.

A path is a simple graph—a graph with no parallel edges (two or more edges between the same two vertices) or loops—whose vertices can be arranged in a linear sequence in such a way that two vertices are adjacent if they are consecutive in the sequence, and are nonadjacent if they are not [7]. Likewise, a cycle is a path in which the first and last vertices of the path are the same, i.e., the nodes in a cycle are arranged in a cyclic sequence. A simple path is a path that does not intersect itself.

4.1.1 The Overlap Graph Approach

The sequence assembly problem can be modelled as a graph problem by making an overlap graph of reads. In the overlap graph, reads are presented as nodes, and the existing overlap between two reads is presented as an edge between corresponding nodes. A modified version of the Smith-Waterman [8] dynamic programming algorithm is usually used to find overlapping reads in almost all assemblers [9].

In an overlap graph, assembling the reads into the genome is equivalent to finding a Hamiltonian path, which is a path that visits every node of the graph exactly once. Unfortunately, finding a Hamiltonian path is an NP-complete problem, and cannot be done in polynomial time. Figure 4.3 shows an overlap graph of a set of reads.

4.1.2 De Bruijn Graph Approach

In 1989 Pevzner [10] proposed a new approach for assembling reads from Sequencing by Hybridization (SBH). Using SBH, a DNA string is reconstructed based on its l-letter substrings. In fact, in hybridization experiments it is determined whether a given query I-mer appears in a target string or not [11]. Therefore,

Fig. 4.3 Overlap graph of
reads. Each read is a node in
this graph and an overlap
between two reads is
presented as an edge in the
graph

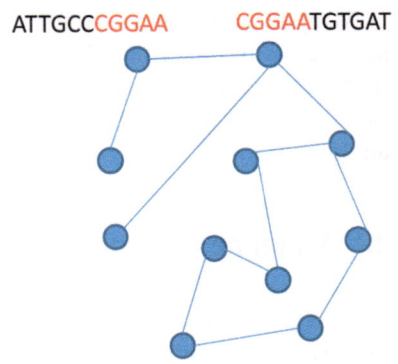

ATTGCCCGGAA   CGGAATGTGAT

SBH reads are short l-tuples. The first approaches to assembling SBH reads were
based on overlap-layout-consensus, which will be explained in the next section, and
using an overlap graph [12, 13]. Pevzner [10] introduced a method based on the
Eulerian path approach for assembling SBH reads. In 2001 Pevzner et al. used this
approach to solve assembly problems. In the new approach, reads are cut into
smaller but regular pieces, called l-mers, and a de Bruijn graph is made from those
l-mers. Using this approach, the complicated step of finding all overlaps between
reads to form an overlap graph is no longer needed, and instead the NP-complete
Hamiltonian path is converted to find a Eulerian path in a de Bruijn graph. There are
polynomial time algorithms for finding Eulerian path problems. However, in
practice, several Eulerian paths can exist in de Bruijn graphs, and finding the
shortest Eulerian superpath is still NP-hard [14]; algorithms use heuristic methods
to compute this super path by applying some modifications to the Eulerian graph.
Another advantage of reducing the fragment assembly to a de Bruijn graph is a
simplification in resolving repeats; as a simple example, Figure 4.4 [15] compares
repeats in an overlap and de Bruijn graph.

A k-dimensional de Bruijn graph is a directed graph whose nodes are all
possible length-k sequences of m symbols. It is clear that each k-dimensional de
Bruijn graph of m symbols has m^k vertices. A de Bruijn graph is a representation
based on all k-mers (length k words), which makes it suitable for high-coverage
very short-read data.

An edge in de Bruijn graphs connects two vertices (k-mers), if one vertices
postfix of length k-1 is equal to the prefix of the other one with the same length.
The edge is directed, and the direction is from the k-mer, including the postfix to
the k-mer including the prefix.

Having a dataset of reads, one can make a de Bruijn graph of it. For example, if
we have an input set (AAGACTC, ACTCCGACTG, ACTGGGAC, GGACTTT),
the list of all 3-mers is (AAG, AGA, GAC, ACT, CTC, TCC, CCG, CGA, CGA,
CTG, TGG, GGG, GGA, CTT, TTT). To create a de Bruijn graph, it is enough to
put the directed edges in the graph according to the input set. The de Bruijn graph
for this set is shown in Fig. 4.5.

Fig. 4.4 a DNA sequence
with a triple repeat *R*; **b** The
layout graph; **c** Construction
of the de Bruijn graph by
gluing repeats; **d** The de
Bruijn graph

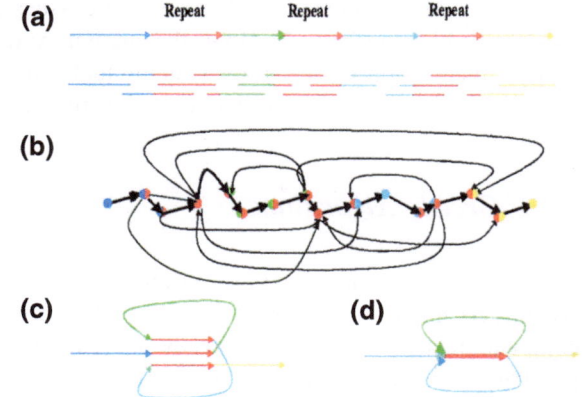

Fig. 4.5   The de Bruijn graph
of an input set (AAGACTC,
ACTCCGACTG,
ACTGGGAC, GGACTTT)

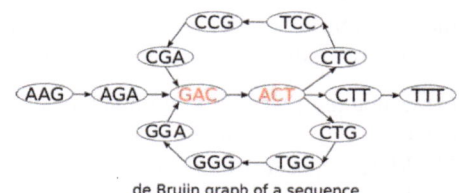

In de Bruijn graph approach assembly algorithms, the graph of input reads are
created and then paths in this graph are used to detect contigs. Finding Eularian
paths is the key to find contigs in this step. Optionally, the algorithm may use other
data—such as paired-end data—in order to make longer contigs and complete the
assembly process. The need for predefined k value, and also errors in reads that
lead to a complex graph structure, are of issues in de Bruijn graph-based assembly
algorithms.

In the next section, main categories of de novo assembly algorithms are
introduced and details of applying graph algorithms into an assembly problem are
given in some selected existing algorithms. The idea behind each of these methods
is explained in the following subsections. Examples of existing algorithms for each
category are also given in each section.

4.2 Classification of De Novo Assembly Algorithms

Existing de novo sequence assembly algorithms can be categorized in three branches: greedy algorithms, overlap layout consensus (OLC) algorithms that use an
overlap graph, and de Burijn graph algorithms that use a de Bruijn (k-mer) graph.
Greedy methods use greedy read extension in order to assemble sequences. In the
following subsections, each of these methods is described in more detail and, some
examples are presented for each. These selected examples show how a general

strategy can be customized to address specific issues in assembly. There are many other assembly algorithms and tools that are not explained here. In Sect. 4.3 of this chapter, there is a list of existing assembly tools for each algorithmic class as well as a comparison of those tools based on the study in [16].

4.2.1 Greedy Algorithms

The shotgun sequence assembly problem was first formalized by finding the shortest common superstring of the set of all reads [17]. Since this algorithm was computationally NP-complete, greedy approaches were introduced to solve the problem. The greedy approach uses a greedy idea—that is, to merge two reads with maximum overlap score at the time (Fig. 4.6). Reads and overlaps are considered to be nodes of graph and edges between them respectively in a graph. Now the problem is reduced to finding a Hamiltonian path in the graph.

Greedy algorithms for sequence assembly can be written in the following steps:

(1) Calculate pairwise alignments of all fragments.
(2) Choose two fragments with the largest overlap.
(3) Merge chosen fragments.
(4) Repeat steps 2 and 3 until only one fragment is left.

The main problem of this approach is the same as all greedy algorithms, i.e., getting stuck in local maxima. A local maxima can occur if the current contig takes on reads that would help further contigs grow even larger.

Examples of algorithms using a greedy approach are PE-Assembler [18], SSAKE [19], SHARCGS [20], and VCAKE [21]. In the following subsections,

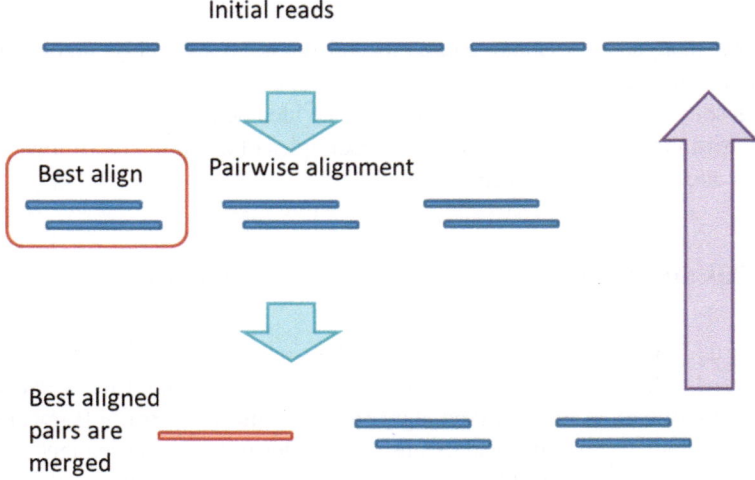

Fig. 4.6 The main steps in greedy algorithms for genome assembly

SSSAKE, SHARCGS, and PE-Assembler algorithms are briefly explained (in more detail).

4.2.1.1 SSAKE

SSAKE—Short Sequence Assembly by progressive K-mer search and $3'$ read Extension—is the first short-read assembler. It is designed for doing the sequence assembly task using unpaired short-reads of uniform length. This algorithm does not use a graph explicitly but its general idea is the same as that of greedy algorithms. SSAKE exists as a program for sequence assembly tasks, which is extended to also make use of paired-end reads and imperfectly matching reads. Instead of using a graph structure, SSAKE stores reads in a lookup table indexed by their prefixes. SSAKE program cycles through sequence data stored in a hash table, and progressively searches through a prefix tree for the longest possible k-mer between any two sequences.

The algorithm iteratively searches for reads that overlap one contig end and chooses candidates that have prefix-to-suffix overlaps whose length is above a threshold. At this point, the program merges two reads with the longest overlap. In the case where there are multiple reads with equally long overlaps, SSAKE chooses reads with end-to-end confirmation in other reads (this favors error-free reads). If there are branches in the merging task, the algorithm terminates. Finally, when there are no reads satisfying the threshold, the program decreases the threshold until a second one is reached (that can be used in previous order). The process is repeated until no more reads exist. SSAKE is available as software at ⟨http://www.bcgsc.ca/bioinfo/software/ssake⟩.

4.2.1.2 SHARCGS

SHARCGS algorithms can be summarized in three steps:, filtering to remove errors, assembly to generate contigs, and, finally, a contig merging step. The SHARCGS algorithm is designed to operate in uniform-length, high-coverage, unpaired short-reads. This algorithm adds pre- and post-processing functionality to the basic SSAKE algorithm. SSAKE is vulnerable to the existence of read errors, and SHARCGS avoids this by using a preprocessing step. Many of the assemblers use this idea and add pre- and post-processing functionality in their assembly process. It helps the assembler to limit the range of errors and ambiguities in input and also omits some errors from the output.

In the preprocessing step, SHARCGS filters erroneous reads by requiring a minimum number of full-length exact matches in other reads. The idea behind this filtering is that reads that have only a few matches in other reads are more likely to be erroneous because the read set is gained from a high-coverage sequencing step. Another criterion considered in the filtering step is the existence of overlapping partners that can be a property of a correct read. A minimum overlap size for

(a)

Core **assembly algorithm**:
1. Select a yet unused read to nucleate a new contig C
2. Elongate C at its 3' end
    2.1. Set n to r
    2.2. Find available read R with prefix P of length n matching the end of C, decreasing n above o_{min}.
    2.3. Use suffix of R as potential extension E
    2.4. Join last r nucleotides of C to E to form check region M.
    2.5. Generate all substrings S of length o_{min} from M and its reverse complement
    2.6. Match reads with prefixes in S to M or its reverse complement, respectively
    2.7. If all reads match, extend C by E, remove R from available reads, and continue with step 2.1.
3. Compute reverse complement of C and redo step 2.

(b)

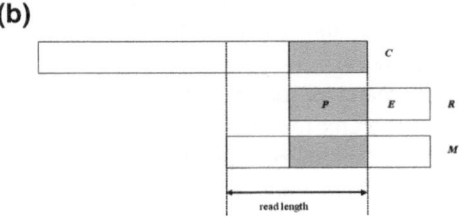

Fig. 4.7   (from [20]). Description of the core assembly algorithm. **a** Pseudocode overview of the steps during assembly of a single contig. The parameter o_{min} controls the stringency of the algorithm, and r denotes the read length. **b** Illustration of the elongation step. Contig$C$ is to be elongated to the *right*. Read R is a candidate for elongation found in the dataset of reads, because its prefix (*gray*) matches the end of C perfectly. The suffix of read R (*white*) is the potential extension E for contig$C$. The length of the check region M is the sum of read length r, and the length of the extension E. Substrings of M and its reverse complement are used to search for matching read prefixes in the dataset. Only if all of these reads match M exactly is C extended by E

considering other reads as overlapping partners is a parameter in the filtering step and is chosen to be at least half of the read length. An optional filtering step is also applied in preprocessing, which is checking whether the combined QVs of matching reads exceed a minimum threshold. Combining QVs is done using the method in Phrap [22]. Finally, one copy of each read is kept in the read set used in the core algorithm. The SHARCGS's core algorithm is shown in Fig. 4.7.

The core SHARCGS algorithm is based on using a prefix-tree look-up for contig extension. A contig is extended as long as there are reads with a prefix of minimal length, which overlap its end perfectly. In extending a contig, the algorithm may face some ambiguities which stop it. For example, if two contigs are extended (they have a different first part), but because of ending the last parts in repeated areas, two contigs merge at the end. In this case, the elongation of contigs is terminated. Also in cases where there are two possible ways to extend a given contig, and there is no way to resolve ambiguity, the algorithm stops. After contig extension at the 3' end is terminated, it computes its reverse complement and tries elongation at the other way in a similar manner. The elongation process is presented in Fig. 4.7b. In order to reduce the effect of ambiguities, SHARCGS first tries to search for possible ambiguities.

SHARCGS filters the raw read set three times, each at a different stringency setting, to generate three filtered sets. Very stringent levels lead to short contigs because too many positions in the target sequence are filtered due to errors, and low stringency levels produce long contigs. Finally it assembles each set separately by iterative contig extension. In the post-processing, the three contig sets are merged using sequence alignment. SHARCGS is available at http://sharcgs.molgen.mpg.de/

.

4.2.1.3 PE-Assembler

PE-Assembler is another greedy assembler for assembling short reads. Instead of using de Bruijn graph approaches (like most of recent assemblers), the PE-Assembler uses the simple greedy approach similar to SSAKE, VCAKE and SHARCGS. It also uses paired-end reads for resolving ambiguity. This method is capable of handling large datasets and producing highly contiguous and accurate assemblies compared to existing methods before it within a reasonable amount of time [18]. The PE-Assembler can be found as a software package at http://www.comp.nus.edu.sg/~bioinfo/peasm/.

There is a simple idea behind the design of PE-Assembler: to use paired-end data for contig extension. The algorithm reduces ambiguities in the read extension phase using paired-end reads. For resolving ambiguities in the case where there are two possible reads to extract the current contig, the algorithm uses the one where its other end matches the extension's other end. This method is shown in Fig. 4.8.

PE-Assembler consists of five main steps: (1) read screening; (2) seed-building; (3) contig extension; (4) scaffolding; and (5) gap-filling. The steps are described in the following subsections.

Read screening A read screening step is defined to identify "solid" reads. Solid reads are reads from an actual genome (i.e., error-free and non-repetitive reads

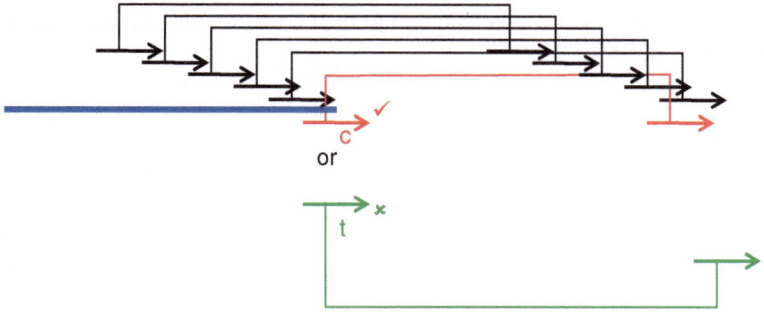

Fig. 4.8 Resolving ambiguities in read extension using paired-reads in PE-Assembler [18]

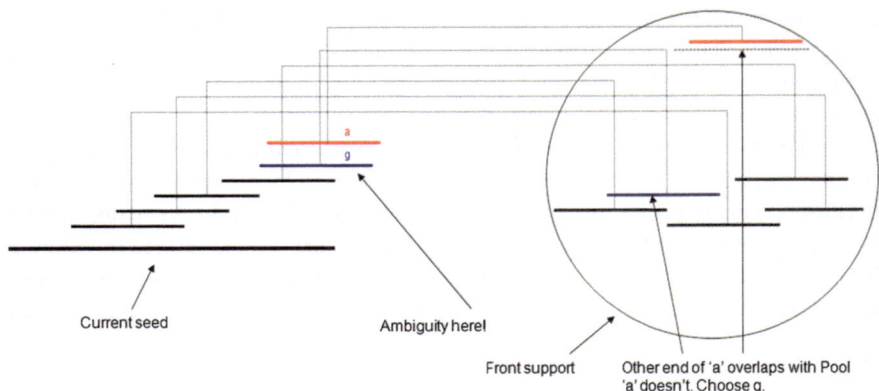

Fig. 4.9 Resolving multiple feasible extensions while seed-building, using paired-end reads [18]

from a genome sequencing step). In this step, filtering the input for omitting erroneous reads is done by statistical analysis (Fig. 3.4). The idea behind the analysis is:

- If a k-mer occurs once, it is likely to be a sequencing error.
- If a k-mer occurs too many times, it is likely to be a repeat.

The area between two determined spots in the diagram shows the interval at which solid reads exist. The solid reads are starting points for extension. This read screening method is the same as what was introduced in [15].

Seed-building A seed is a contiguous region that is the result of read extension and whose length is at least MaxSpan. Starting from some "solid" reads, PE-Assembler extends the read from both 5' and 3' ends. If there are multiple feasible extensions, mates are used to resolve ambiguity. In Fig. 4.9, g has support while a does not have support. Hence, g is correct.

Since this process does not handle ambiguities that are due to sequencing errors, an extra step is needed. In this case, every candidate is extended base up to a distance of *ReadLength,* and wrong paths (generated from erroneous reads) will be terminated prematurely and therefore detected. A path that is extended in a single strand and is not branched is considered to be a correct path. An example of this is shown in Fig. 4.10.

Fig. 4.10 Resolving ambiguities due to a sequencing error in PE-Assembler [18]

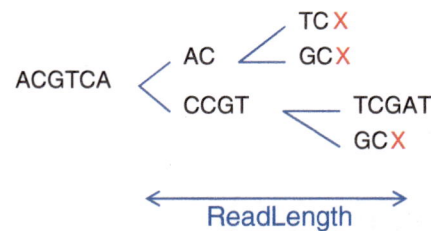

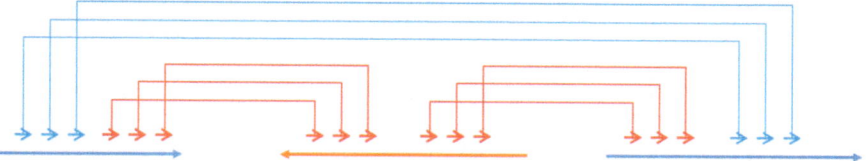

Fig. 4.11 Ordering contigs in scaffolding step in a PE-Assembler [18]

Finally, seeds are verified by checking if at least one paired-end read overlaps with the 3′ and 5′ end of it. Unverified seeds are ignored in the rest of the assembly process.

Contig extension Contig extension is performed to extend verified seeds to form longer contigs. Seeds are extended using paired-end reads in this step—in contrast to previous steps, in which single reads were used for extension. It is feasible since seeds are longer than maxSpan. After no paired-end reads were available for the extension, the algorithm looks for possible reads again. Meanwhile, repeats and errors are handled as in previous steps.

Scaffolding problem Having a set of contigs of some genome X and a set of DNA-PETs of some genome X as input, the aim of this step is to find the correct ordering and orientation of the contigs. Figure 4.11 shows an example of the scaffolding process using paired-read data. Paired-end reads that are mapped to repeats are discarded in this step since they may lead to incorrect mappings. After finding proper ordering between contigs, the contig graph is built (Fig. 4.12).

Gap-filling After the scaffolding step, there are still gaps remaining in the assembled sequences. These gaps are usually within repeat areas, since reads within this area were discarded in earlier steps. Using paired-end data, gaps can be filled in the similar extension manner, but this time using both 3′ and 5′ ends in order to be consistent with both contigs around the gap (Fig. 4.13).

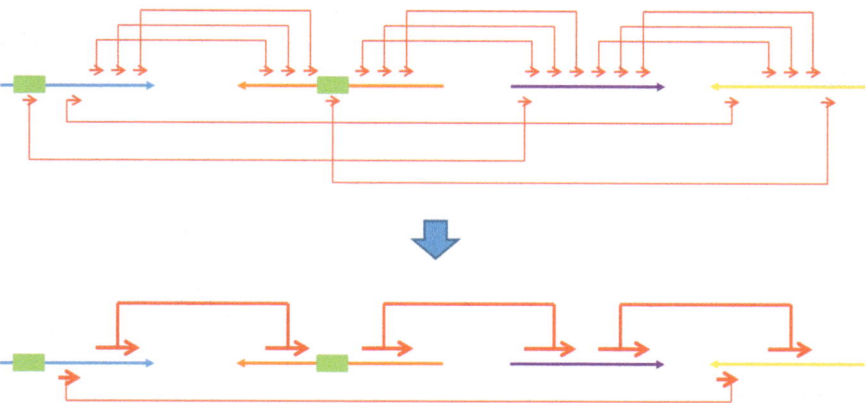

Fig. 4.12 Building a contig graph from ordered contigs. Paired-end reads within the repeat area—shown in *green*—are ignored in this step [18]

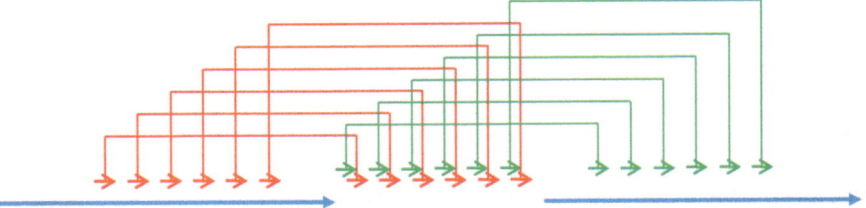

Fig. 4.13 Gap-filling step in a PE-Assembler [18]

PE-Assembler authors discuss the issue of parallelization in the algorithm, and claim that almost all the algorithm steps can be improved by parallelization. Read screening, which is largely disk- bound and its parallelization does not improve runtime, and the actual scaffolding step, which is not time consuming, are carried on a single thread.

4.2.2 Overlap Layout Consensus (OLC) Algorithms

Overlap layout consensus methods are based on graph theory. In this method, an overlap graph is built from reads and the assembly problem is reduced to finding a Hamiltonian path—a path that contains each node exactly once—in the graph. ARACHNE [23], Celera [1] and its revised version for short-reads [24], CAP3 [25], and Newbler [26] are assemblers that use this method as their core idea.

An OLC algorithm starts by finding overlaps between reads (or graph nodes). In fact, it must check possible overlaps between any two reads in the input read set. The layout step will simplify the overlap graph by removing redundant information and will put these reads together using identified overlaps. The final step is finding a consensus for the existing layout (Fig. 4.14). The overlap step is computationally very expensive and therefore this approach is more suitable for whole-genome shotgun sequencing reads gained from Sanger technology. Also, the Hamiltonian path problem is an NP-complete problem in itself, needing heuristic solutions.

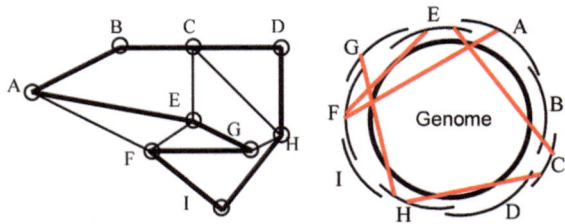

Fig. 4.14 Overlap graph for a bacterial genome. The thick edges in the picture on the *left* (a Hamiltonian cycle) correspond to the correct layout of the reads along the genome (figure on the *right*). The remaining edges represent false overlaps induced by repeats (exemplified by the *red lines* in the figure on the *right*). [http://www.cbcb.umd.edu/research/assembly_primer.shtml]

4.2.2.1 Celera and CABOG

The Celera [1] algorithm was introduced for the whole-genome assembly of *Drosophila*. A challenge that Celera tries to overcome is handling repetitive parts in the genome that may mislead the assembly. For this purpose it uses mate-pair information data to resolve problems in repeated areas of the genome. Celera used the OLC method and was mainly developed to use the Sanger shotgun sequencing data; it was later revised to use NGS data. The revised algorithm is called CABOG (Celera Assembler with the Best Overlap Graph) [24]. The Celera algorithm is designed as a pipeline, as shown in Fig. 4.15.

The first step of this pipeline is the screener, which selects high-quality data. In this step, input fragments are compared to a repeat database to see if they match a known repetitive element in the genome. In the case of a match, the read is omitted in further steps of the pipeline. In the special case of *Drosophila*, the algorithms use an existing repeat library. In the next step, the overlapper, a BLAST-like algorithm searches for overlaps—in order to assemble reads in the OLC manner— and chooses sequences with less than a 6 % difference for merging. In the Unitigger step, unitigs are formed. Unitigs are made from the assembly of fragments whose arrangement is uncontested by overlaps from other fragments. Unitigs that

Fig. 4.15 In Celera's assembly pipeline, sequences flow from one stage to the next. Each stage performs work on its input stream, producing a stream of outputs reflecting its transformational function [1]

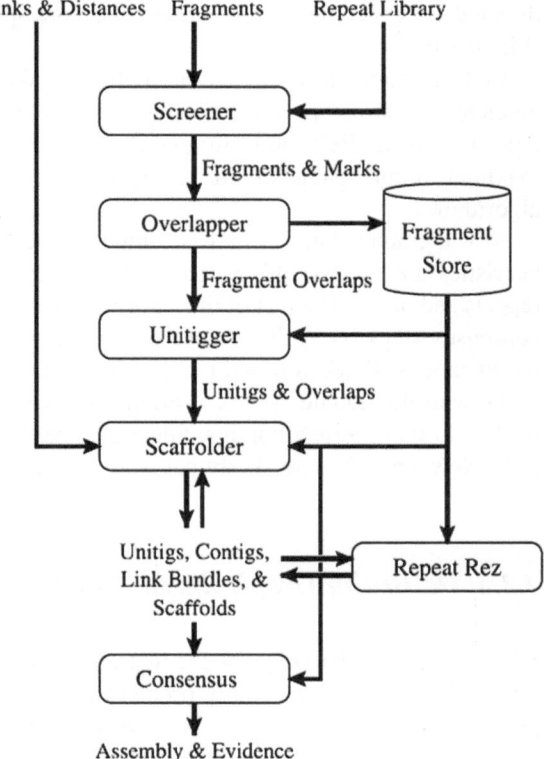

present unique DNA are called U-unitigs; they are extended, using repeat sequences, where their tips are matched to one. These are repeat boundaries. Some repetitive overlaps between unitigs are identified and removed, using these repeat boundaries. Finally, in the scaffolding step, pairs of mates or BAC ends are used to confirm the linking of U-unitigs and forming of scaffolds. The output of this step is corrected using mate-pair data—a repeat resolution module—and the final consensus is created during an iterative process.

The CABOG assembler uses the idea that different types of input data—like reads from Sanger and pyrosequencing platforms—can be used together for the assembly. Because of differences in accuracy, coverage, average read length, and also the availability of paired-end protocols, it is not straightforward to use an arbitrary assembly algorithm for hybrid data, and this may lead to a poor performance of the algorithm [24].

CABOG reuses the Celera Assembler scaffold (modified to recover trimmed base calls) and consensus modules (modified to specify alternate consensus sequences in polymorphic regions). It also uses a base call correction method introduced in ARACHNE [23]. In this step, it filters out some erroneous reads. Then it selects the "best" overlaps that survive this filter, and it also applies a filter for a minimum length of the alignment. An overlap graph is generated from these best overlaps (overlaps from reads with containment overlaps are disregarded). This graph is called Best Overlap Graph (BOG), in which there exists at most one directed edge per node representing the corresponding read end's best overlap (Fig. 4.16).

BOG is implemented using as multiple linked lists in an array of reads, an efficient data structure,resulting in an extreme data reduction of the sets of overlaps. Cycles in BOG are eliminated by deletion of arbitrary edges, and then maximal simple paths in this graph are used to build unitigs with a greedy algorithm.

Now a graph is built from these unitigs using some paired end constraint. Some heuristics are used in this part of the algorithm in order to deal with genomic repeats and noise. The rest of the algorithm, i.e. contig generation and scaffold and consensus steps, is again a reuse of Celera Assembler pipeline without special modifications. Reads rejected previously due to constraint violations, may be used in the scaffold module. Mate constraints also can be used here to match individual reads into their correct contigs. Some of other OLC based algorithms like Edena [27], Newbler [26], and shorty software [28] are briefly introduced in [29].

4.2.3 De Bruijn Graph-Based Algorithms

The Euler assembler [15] was the first algorithm that used the de Bruijn approach for solving sequence assembly problems. Velvet [30], Euler-USR [31], AllPaths [32], Abyss [3], and IDBA [33] are some other assembly algorithms that use this

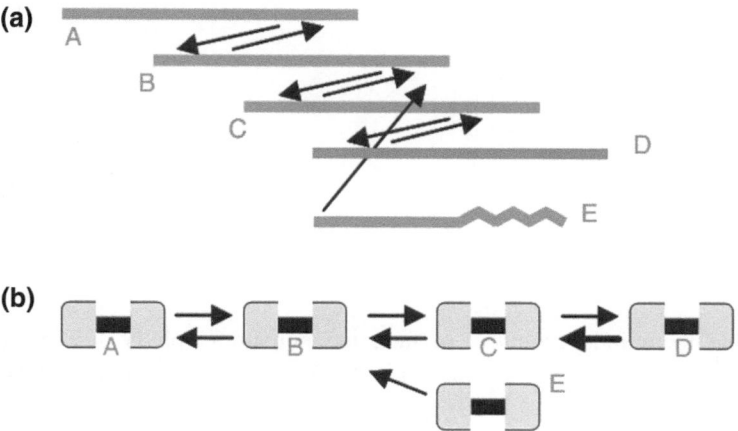

Fig. 4.16 Two representations of a best overlap graph. In **a**, the layout resembles a multiple sequence alignment. In **b** each read is represented by two nodes joined by an undirected edge. Arrows represent best overlaps, where best means covering the most sequence. There are mutual best overlaps between successive pairs of reads A through D. Due to erroneous bases at one end (wavy line), read E has a non-mutual best overlap to B. Paths span undirected and directed edges alternately. Path EBA converges on path ABCD. CABOG scores read E lower than the others since only three reads are on paths from it. Starting with any one of the high-scoring reads, CABOG would build initial unitig ABCD, then E. Using saved information about each path intersection, CABOG would discount the intersection at B because the path from E spanned only one read before B. It would break ABCD only if there were also a change in read arrival rate at B, which is not the case here. Although linear-time directed-path following finds the longest possible unitig in this constructed case, it is not guaranteed to do so whenpaths span multiple intersections [24]

approach. In this section we describe the Velvet [30], Euler-USR [31], and All-Paths [32] algorithms.

4.2.3.1 AllPaths

ALLPATHS is an algorithm for short-read sequence assembly. This algorithm was published both on simulated [32] and real data, ALLPATHS2 [34]. ALLPATHS applies error-corrections based on the method used in [15] on reads. Each read is kept, edited or discarded in this phase, in which reads with high frequency and high quality are considered to be trusted reads for next steps. It then stores reads in a compact, searchable data structure. This structure is used in order to avoid the process of overlap finding. Instead, reads are searched in this data structure for finding matches and merge reads. The algorithm generates a unipath graph from reads and then localizes read sequences before assembly. A unipath (Fig. 4.17) is a maximal unbranched sequence that is retrieved for a given minimum overlap k in the given genome. Each k-mer can exist in one unipath generated from the reads. Localization is a way of using pairs to isolate small regions of the genome and to

Fig. 4.17 A unipath is a
maximal unbranched
sequence generated from
consequence reads [32]

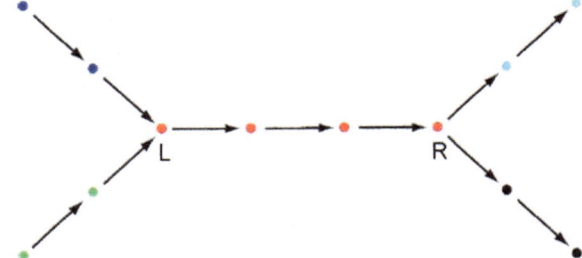

assemble them independently. In fact, the unipath graph is the best possible
assembly of the genome from reads of length k in theory given infinite coverage by
perfect reads [32].

Next, AllPaths assigns numbers to k-mers in the reads and stores them in a
searchable database. Unipaths are constructed by a walking process on the reads in
the database until a branch is reached, For read pairs, the assembly process is more
complicated. It tries to find a path from one end of the read pair to the other end by
filling the gap with high-coverage reads. Assemblies of genomic regions are
formed around seeds (ideally, long unipaths with low copy numbers). For each
unipath, closest unipaths to its left and right side are computed, and if the distance
between the two are less than 4 kb, the middle unipath is removed. After all such
unipaths are removed, the remaining ones form the seed unipaths.

In the next step, neighborhoods around seeds are assembled. A neighborhood is
defined as one seed plus 10 kb on each side of it. In this step a collection of low-
copy number unipaths are defined, using iterative linking (Fig. 4.18a), and then
two sets of read clouds are constructed: primary, which include only reads whose
true genomic locations are near seed (in detail it contains those reads incident upon
one of the neighborhood unipaths), plus their partners. Some of them reach into
gaps (Fig. 4.18b), and secondary, which contain all short-fragment read- pairs—
about 0.5 kb—near the seed. Because too many closures may exist in this step, a
short-fragment pair merger is used to progressively merge the secondary read
cloud pairs.

After that, all closures of all merged short-fragment pairs are computed, and all
paths are generated. Local assemblies are merged together by iteratively joining

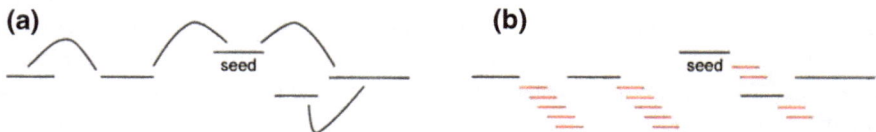

Fig. 4.18 [32] Localization. **a** *Lines* represent unipaths, and *curves* represent paired-read links
between them; from seed, iteratively linked to low-copy-number unipaths within a 10-kb radius
of it. **b** Reads aligning to these unipaths have partners (*red*) that dangle in repetitive gaps between
them

closures, and finally global assembly is built, using all of these local assemblies. To remove detritus, eliminate ambiguity, and pull apart regions where repeats are assembled on top of each other, a final editing step is performed on the final graph.

4.2.3.2 Velvet

Velvet is an algorithm for de novo assembly, which is suitable for short-read assembly (25–50 bp). Applying Velvet to very short reads and paired-ends information only can produce contigs of significant length, up to 50-kb N50 long in simulations of prokaryotic data and 3-kb N50 on simulated mammalian BACs [30]. Velvet algorithms can be described in the following steps:

Constructing a k-mer hash table In order to store k-mers, Velvet uses a hash table approach. A hash table is a data structure that maps values as keys (for example a k-mer) to their associated values (an ID of a read containing that k-mer). A hash function is used to transform the key into the value index in the hash table.

For a predefined value of k, usually $k = 21$ for 25-bp reads, the hash table stores the ID of the first read containing that k-mer and also the position of the k-mer in that read. Each k-mer is recorded simultaneously to its reverse complement. A second database is also created, which shows, for each read which of its original k-mers are overlapped by subsequent reads.

Building a de Bruijn graph Just like what was explained for general de Bruijn graph approaches, nodes are k-mers, and an edge exists between two nodes if the (k-1) suffix of a node equals the (k-1) prefix of a node, add a directional edge between them. In Velvet's simplified graph, nodes are called "block." Each block is a representation of some overlapping reads (k-mers having k overlapping nucleotides). The block contains the last nucleotides of related reads and an edge exists between two blocks if the last read of the first block overlaps k-1 bases with the first read of the second block. Figure 4.19 shows an example of block structure used in Velvet that clearly explains the use of the database design.

Simplification of the graph After the graph is constructed, some simplification is applied to it in order to remove fragmentations in the graph structure (fragmentation is the source memory usage). The simplification uses a simple idea: whenever a node A has only one outgoing arc that points to another node B that has only one ingoing arc, the two nodes are merged (Fig. 4.20).

Error-correction There are some error-correction steps in Velvet that are applied to the graph in this step:

(1) Error-removal: errors can be due to the sequencing process. It is important to distinguish between errors and polymorphism. In Velvet, this can be done using the expected coverage of genuine sequences. All low coverage sequences are considered to be errors in this step, and others are considered to be polymorphisms.

Fig. 4.19 Example of block representation in Velvet [30]

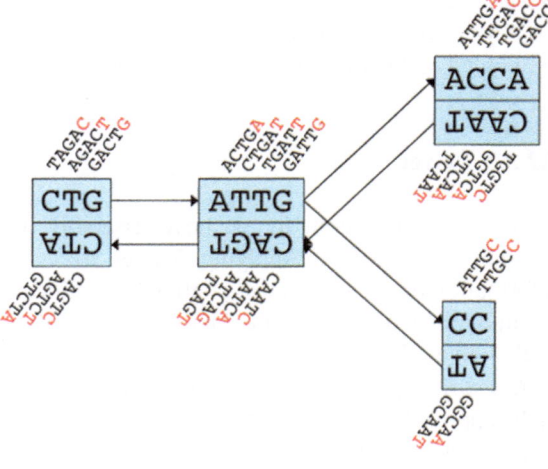

Fig. 4.20 Merging two nodes having just one outgoing and one incoming edge in order [30]

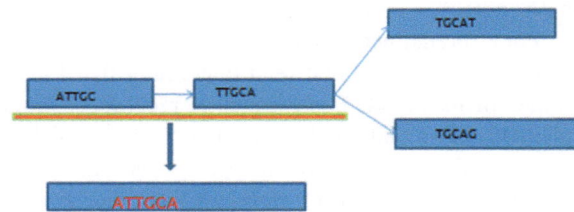

(2) Removing tips: tips exist due to errors at the edges of reads. A tip is a path that is disconnected at one end and therefore more likely to be an error. In the tip-removing step, all tips shorter than 2 k are removed.

Removing bubbles with the Tour Bus Algorithm Two paths are considered redundant if they begin and end at the same nodes (forming a "bubble") and contain similar sequences. Such bubbles can be created by errors or biological variants, such as SNPs or cloning artifacts prior to sequencing. Erroneous bubbles are removed by an algorithm called "Tour Bus," an example of which, as used in Velvet, is shown in Fig. 4.21.

In this step, the graph is preprocessed and ready for the final process, finding the Eulerian path. A Eulerian path in the graph is onr that visits every edge at least once. The final Eulerian path that is found in the graph is considered to be the assembly result.

As described earlier, short-read assembly is problematic with the repeated structure of the genome being sequenced. The Velvet algorithm proposes a module called "Breadcrumb" to make use of pair-end reads information to resolve repeat areas and better merging of contigs.

Fig. 4.21 Example of Tour
Bus error correction.
a Original graph. **b** The
search starts from A and
spreads toward the *right*. The
progression of the *top* path
(through B′ and C′) is stopped
because D was previously
visited. The nucleotide
sequences corresponding to
the alternate paths B′C′ and
BC are extracted from the
graph, aligned, and
compared. **c** The two paths
are judged to be similar, so
the longer one, B′C′, is
merged into the shorter one,
BC. The merger is directed
by the alignment of the
consensus sequences,
indicated in *red lines* in B.
Note that node X, which was
connected to node B′, is now
connected to node B. The
search progresses, and the
bottom path (through C′ and
D′) arrives second in E. Once
again, the corresponding
paths, C′D′ and CD are
compared. **d** CD and C′D′ are
judged to be similar enough,
and the longer path is merged
into the shorter one [30]

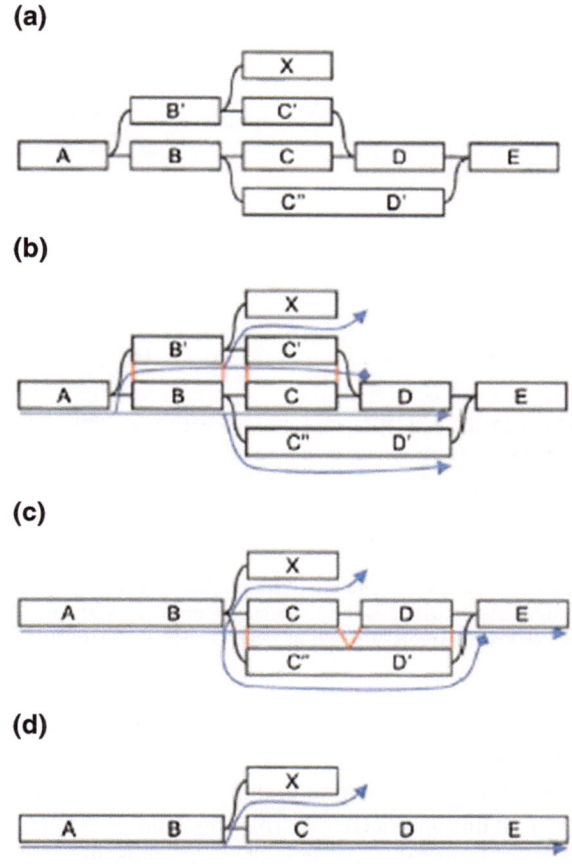

4.2.3.3 Euler-Usr

The EULER-USR algorithm is based on the notion of repeat graphs. A repeat
graph of a genome (or reads) is a simplified version of the de Burijn graph with
small bulges and whirls removed (Fig. 4.22e, f). The key point in this algorithm is
based on this observation, that a repeat graph of a whole genome can be
approximated from a repeat graph generated from reads. Reads may be corrected
by mapping them to repeat graphs of the genome, if the graph is known. The idea
of EULER-USR is to construct repeat graph from accurate reads, and then map the
entire read set (with inaccurate prefixes) to this graph by threading.

EULER-USR consists of three main steps that are explained in this section:

(1) Detecting the set of accurate reads and trying to improve their accuracy using
     high-frequency k-mers.
(2) Constructing the repeat graph on error-corrected prefixes using k-mers.
(3) Simplifying the repeat graph after transforming mate-pairs into mate-reads.

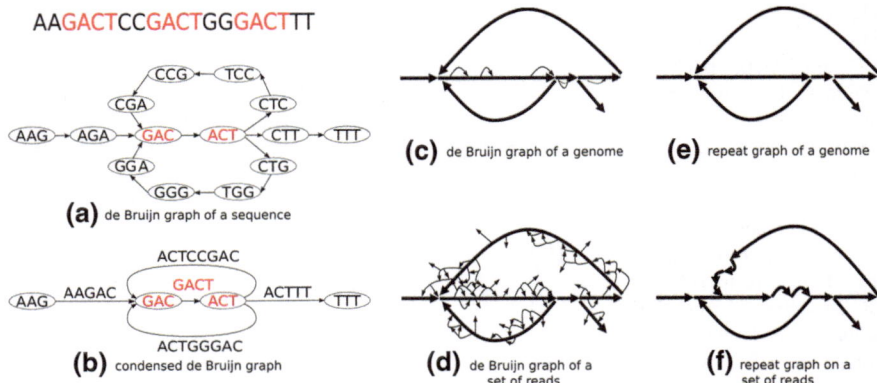

Fig. 4.22 From de Bruijn graphs to repeat graphs. The de Bruijn graph of a sequence contains a vertex for every k-mer in the sequence, and an edge (u, v) for every pair of consecutive (overlapping) k-mers in the sequence (**a**). The condensed de Bruijn graph replaces all paths containing nonbranching vertices by a single edge labeled by the sequence that generated the path (**b**). When the condensed de Bruijn graph is constructed on a genome, it contains some small bulges and whirls representing repeats with slightly varying repeat copies (**c**). In the repeat graph, the bulges and whirls are removed (**e**). The de Bruijn graph of reads contains additional spurious bulges and whirls caused by sequencing errors in reads (**d**). The goal of the Eulerian assembly is to construct the repeat graph of reads (**f**) that approximates the repeat graph of the genome [31]

In the error-correction phase, the algorithm tries to detect the longest read prefix of Illumina reads that may be error corrected, and to discard the reads that cannot be corrected. This phase is done using the Spectral alignment (SA) algorithm in [15]. In this approach, the error-free read is a read where all of its k-mers are solid. A solid k-mer is a k-mer with a minimum threshold of appearance in the k-mers from the read set. Then a greedy approach is used to find the minimum number of mutations that make every k-mer in a read solid. This operation is continued until there are no mutations found, or the k-mers are made solid, and this phase of the algorithm outputs the solid prefixes of reads.

The de Burijn graph is now made from error-corrected reads, but it is not error-free. For example, there are still some errors or SNPs in the reads that create short undirected cycles called "bulges" in the graph. In addition, reads may contain errors at their endpoints, which result in erroneous sources or sinks. There may also be some chimerical reads that may connect two unrelated contigs. Transforming a de Burijn graph to a repeat graph necessitates resolving all of these errors [35].

The next step is the simplification of a repeat graph by transforming mate-pairs to mate-reads. This transformation is practical if a single path can be found from read$_{start}$ to read$_{end}$. The gap- filling module defined in this step is a modification of the EULER-DB method [15]. The idea behind this step is that if several paths between read$_{start}$ and read$_{end}$ are found, the path with the maximum support of mate-pairs is selected, but only if the support is above a predefined minimum threshold.

Finally, error-prone reads are assembled. At this phase, first reads are corrected using a repeat graph. Since repeat graphs are made of prefixes, each read can be corrected by processing all sub-paths continuing that read in the repeat graph. This process is referred to as "threading" in the repeat graph. Threading is done in five steps: (1) detecting accurate reads on the repeat graph and generating extremely accurate reads; (2) constructing the repeat graph on error-corrected k-mers, which generates the set of k-mer contigs; (3) threading entire reads through the repeat graph to extend the effective read length and generating threaded reads; (4) constructing the repeat graph on threaded reads and generating l-mer contigs (l > k); and (5) simplifying the repeat graph by transforming mate-pairs to mate-reads. In this final assembly, repeats of length l and shorter are resolved.

4.3   Comparison of Algorithms

As mentioned in Sect. 3.6, in order to evaluate assembly algorithms, it is not easy to compare assembly programs. There are no unique evaluation criteria (for example, a quantitative comparable measure) that specifies the quality of assembly output as a whole and can be compared for different assemblers and, as discussed in Sect. 3.6, there is a trade-off between already introduced measures (for example, between accuracy and contiguity measures). Different assemblers may be good at different existing criteria, and this makes the comparison difficult. There are several assemblers for various input data, and each of them uses various parameters or additional information that makes it nontrivial to compare them.

This section gives a general comparison of existing assembly tools used in the work done in [16]. In this paper, a new category of assembly algorithms is introduced, called "Branch and Bound" (B&B). This category is defined based on the method used in the SUTTA [36] assembler. In SUTTA, assembly is viewed as an optimization problem. SUTTA generates a set of all possible "consistent layouts" as feasible solutions and tries to use branching and the bound technique to prune the search space.

Assemblers compared in this review are long-read assemblers including ARACHNE [23], CABOG [24], Euler [15], Minimus [37], PCAP [38], Phrap [22], SUTTA [36], and TIGR [39], as well as short-read assemblers including ABySS [3], Edena [27], Euler-SR [35], SOAPdenovo [2], SSAKE [19], SUTTA [36], Taipan [40], and Velvet [30]. Table 4.1 [16] shows the set of assembly tools, their input read types, algorithms, and given references. In this table, SBH is equivalent to the de Bruijn graph approach, and seed and extend falls into the greedy category due to what has been explained in this book. This category is separated in the references because of the use of prefix-tree data structure in their greedy approach.

In [16], benchmark data is also used to compare the number of contigs, number of big contigs (≥10 kbp), max and mean contigs, N50, and the big contig coverage

Table 4.1 List of sequence assemblers [16]

Name	Read type	Algorithm	Reference
SUTTA	Long & short	B&B	(Narzisi and Mishra 2010, [41])
ARACHNE	Long	OLC	(Batzoglou et al. 2002, [42])
CABOG	Long & short	OLC	(Miller et al. 2008, [43])
Celera	Long	OLC	(Myers et al. 2000, [44])
Edena	Short	OLC	(Hernandez et al. 2008, [45])
Minimus (AMOS)	Long	OLC	(Sommer et al. 2007, [46])
Newbler	Long	OLC	454/Roche
CAP3	Long	Greedy	(Huang and Madan 1999, [47])
PCAP	Long	Greedy	(Huang et al. 2003, [48])
Phrap	Long	Greedy	(Green 1996, [49])
Phusion	Long	Greedy	(Mullikin and Ning 2003, [50])
TIGR	Long	Greedy	(Sutton et al. 1995, [51])
ABySS	Short	SBH	(Simpson et al. 2009, [52])
ALLPATHS	Short	SBH	(Butler et al. 2008/2011, [20, 53])
Euler	Long	SBH	(Pevsner et al. 2001, [54])
Euler-SR	Short	SBH	(Chaisson and Pevzner 2008, [32])
Ray	Long & short	SBH	(Boisvert et al. 2010, [31])
SOAPdenovo	Short	SBH	(Li et al. 2010, [55])
Velvet	Long & short	SBH	(Zerbino and Birney 2008/2009, [15, 56])
PE-Assembler	Short	Seed-and-extend	(Ariyaratne and Sung 2011, [35])
QSRA	Short	Seed-and-extend	(Bryant et al. 2009, [57])
SHARCGS	Short	Seed-and-extend	(Dohm et al. 2007, [58])
SHORTY	Short	Seed-and-extend	(Hossain et al. 2009, [2])
SSAKE	Short	Seed-and-extend	(Warren et al. 2007, [59])
Taipan	Short	Seed-and-extend	(Schmidt et al. 2009, [60])
VCAKE	Short	Seed-and-extend	(Jeck et al. 2007, [61])

Reads are defined as "long" if produced by Sanger technology and "short" if produced by lllumina technology. Note that Velvet was designed for micro-reads (e.g. lllumina) but long reads can be given in input as additional data to resolve repeats in a greedy fashion. doi:10.1371/journal.pone,0019175.t001

percentage is compared on these tools, on single-reads and also with the help of paired-end reads. This kind of study, which is done using the same conditions on specific benchmark data, can be used to evaluate assembly quality measures and compare them on different algorithms. Still, it cannot be said that the result would be the same on a different input dataset.

The benchmark dataset used in this study (presented in Table 4.2) is selected by considering criteria such as reproducibility, accessibility in the public domain, etc. Also, this is considered for the data to include all possible genome structures—for example, variation in read length, coverage and error rate.

The first three sets are bacterial genome Sanger reads: *Brucella suis* [62], *Wolbachia sp.* [63] and *Staphylococcus epidermidis* RP62A [64]; all are available at the NCBI Trace Archive and CBCB website (www.cbcb.umd.edu/research/benchmark.shtml). The Human Y chromosome on the p11.2 region is also selected

Table 4.2 Benchmark data [16]

Genome	Length (bp)	Num. of reads	Avg. read length (bp)	Std. (bp)	Coverage
Brucella suus	3,315,173	36,276	895.8	44.1	9.8
Wolbachia sp.	1,267,782	26,817	981.9	50.6	20.7
Staphylococcus epidermidis	2,616,530	60,761	900.2	46.2	19.9
Chromosome Y*	3,000,000	37,530	800	88	10
Staphylococcus aureus	2,820,462	3,857,879	35	0	47.8
Helicobacter acinonychis	1,553,927	12,288,791	36	0	2S4.6
Escherichia coli	4,639,675	20,816,448	36	0	161.5

First and second columns report the genome name and length; columns 3 to 6 report the statistics of the shotgun projects: number of reads, average and standard deviation of the read length and genome coverage (*region [35,000,001–38,000,000]), doi: 10.1371/journal.pone.0019175.t002

because it has a pathologically complex genome patterning structure (repeats, duplications, indels, head-to-head copies, etc.). For this part of *chromosome Y*, shotgun data is simulated. Also, two mate-pair libraries of size ($\mu = 2500, \sigma = 166$) and ($\mu = 10000, \sigma = 1300$) are generated for 90 % of the reads (45 % from the first and 45 % from the second library, so other reads have no mate). An error rate of 1 % is implemented in each read.

Short-read data used in this study are datasets from the *Staphylococcus aureus* strain MW2 [65] (freely available at the Edena assembler website: www.genomic.ch/edena.php), *Helicobacter acinonychis* strain Sheeba genome [66], presented in SHARCGS [20], available at sharcgs.molgen.mpg.de, and, finally, 20.8 million paired-end 36 bpIllumina reads from a 200 bp insert *Escherichia coli* strain K12 MG1655 [67] library (NCBI Short Read Archive, accession number SRX000429).

Tables 4.3, 4.4 gives the assembly results from various tools for long- and short-reads respectively, without mate-pair data. In the reference, other comparisons exist for mate-pair data as well. As mentioned in Sect. 3.6, a new metric for comparison, named the "Feature-Response Curve" (FRC), is introduced in the reference which captures the quality and contig size trade-off. FRC curves are shown in the reference to give more information about assembly tools other than the common criteria comparisons presented here. As an example of using FRC curves, we can see that Phrap performs better (high coverage and N50) in Table 4.3, while its mis-assemblies within long contigs are not captured by measures like N50. In fact, this tool is not able to handle large-range genome structures.

Figure 4.23 shows FRC curves for the assemblers compared in Table 4.3 for *S. epidermidis* and *Chromosome Y* (P11.2 region) genomes with no mate-pair data. In this figure, the x-axis is the minimum number $\emptyset$ of error/features allowed in the contigs, and the y-axis is the approximate genome coverage achieved by all the contigs, such that the sum of their features is $\leq \emptyset$. Overlapping regions of contigs are double-counted in the coverage given in the figure. This kind of

Table 4.3 Long-read comparison without mate-pairs [16]

Genome	Assembler	# contigs	# big contigs (≥ 10kbp)	Max (kbp)	Mean big contigs (kbp)	N50 (kbp)	Big contigs coverage (%)
Brucella	Euler	280	118	82	22	19	78.4
Suls	Minimus	203	101	89	30	32	93.1
	PCAP	88	62	198	53	80	100.7
	PHRAP	54	23	434	126	199	103.2
	SUTTA	73	53	268	62	79	99.2
	TIGR	108	67	182	48	57	98.8
Staphylococcus	Euler	192	75	78	29	32	85.6
epidermidis	Minimus	425	86	119	10	19	80.7
	PCAP	109	36	179	72	114	100.1
	PHRAP	86	22	357	123	183	103.9
	SUTTA	65	31	249	83	116	99.3
	TIGR	94	38	230	68	100	99.8
Walbachia sp.	Euler	604	0	6	0	1	0
	Minimus	1545	37	16	13	2	40.7
	PCAP	1241	41	64	23	3	77.2
	PHRAP	2253	55	64	22	1.8	98.5
	SUTTA	1089	39	87	26	6	80.8
	TIGR	1080	46	46	20	5	73.6
Human	Euler	60	27	403	107	266	96.7
Chromosome Y	Minimus	850	104	48	18	11	63.1
	PCAP	140	38	239	77	112	98.2
	PHRAP	4	4	1869	764	1869	101 9
	SUTTA	15	10	1020	301	712	100.5
	TIGR	1103	108	51	10	8	63.7

Long reads assembly comparison Without mate-pair information (clone sizes and forward-reverse constraints). First and second columns report the genome and assembler name; columns 3–7 report the contig size statistics, specially: number of contigs, nimber of contigs with size ≥ 10 *kbp*, max contig size, and N50 size (N50 is the largest number L such that the combined length of all contigs of length ≥ L is at least 50 % of the total length of all contigs). Finally column B reports the coverage acheived by the large contigs (≥ 10 *kbp*). Coverage is computed by double-counting overlapping regions of the contigs, when aligned to the genome. doi: 10.1371/journal.pone.0019175.t003

comparison shows that SUTTA performs better than all other methods. According to the authors [16], FRC is suffering from not capturing some types of assembly errors, such as one of the most ones—mis-joints—and just considers several error types with a uniform weighing. Table 4.4 presents results from short-read assemblers.

More details on the comparison with available mate-pairs and discussions on FRC curves is presented in the References section.

Table 4.4 Short-read comparison without mate-pairs [16]

Genome	Assembler	# Correct	# Mis-assembled	N50 [kbp]	Mean [kbp]	Max (kbp)	Coverage (%)
S. aureus	ABySS	928	6	7.8	2.9	32.7	98
(*strain MW2*)	Edena (strict)	1124	0	5.9	2 4	25.7	98
	Edena (nonstrict)	740	16	9.0	3.7	51.8	97
	EULER-SR	669	33	10.1	4.0	37.9	99
	SOAPdenovo	867	25	81	3.1	30.8	97
	SSAKE	2073	378	2.0	1.1	9.7	99
	SUTTA	998	11	6.0	2.6	22.8	97
	Taipan	692	16	11.1	3.9	44.6	98
	Velvet	945	5	7.4	2.8	32.7	97
H. acininychis	ABySS	270	8	13.9	5.4	54.7	98
(*strain Sheeba*)	Edena (strict)	336	0	10.1	4.5	36.9	98
	Edena (nonstrict)	302	1	13.2	4.9	35.0	97
	EULER-SR	730	21	4.3	2 1	18.8	98
	SQAPdenovo	479	21	7.3	3.3	29.8	98
	SSAKE	675	156	3.2	1.8	14. 6	99
	SUTTA	313	9	9.6	4.5	41.3	98
	Taipan	271	0	13.3	5.6	48.6	98
	Velvet	278	2	12.8	5.4	49.5	98

Short reads assembly comparison without mate-pair information. First and second columns report the genome and assembler name; columns 3–7 report the contig size statistics, specifically number of contigs, number of contigs witti size ≥ 10 kbp, max contig size, mean contig size, and N50 size (N50 is the largest number L such that the combined length of all contigs of length $\geq$ L is at least 50 % of the total length of all contigs). Finally column 8 reports the coverage achieved by all the contigs. doi:10.1371journal.pone.001917S.t005

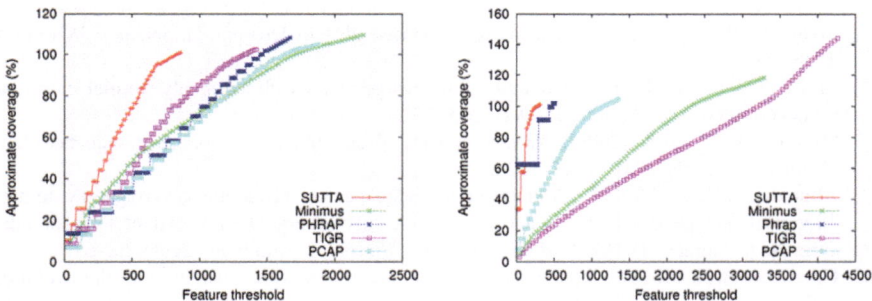

Fig. 4.23 (*Left*) Feature-Response curve comparison for *S. epidermidis* with no mate-pair information; (*right*) Feature-Response curve comparison for *Chromosome Y* (3 Mbp of p11.2 region) with no mate-pair information [16]

4.4 Summary

In this section, the main approaches for solving the assembly algorithm were presented, and for each approach several algorithms were explained. There are three main categories for assembly algorithms: greedy algorithms; overlap-layout-consensus algorithms, which are based on overlap graphs and Hamiltonian path-finding; and de Bruijn graph algorithms, which are based on de Bruijn graphs and Eulerian path-finding in assembly graphs. De Bruijn graph methods show more strength in short-reads and resolving repeats, while overlap graph methods are more suitable for Sanger shotgun data. While it seems that greedy methods were applicable just on long sequences, some tricks used in new algorithms, like the PE-Assembler which because of using paired-end reads, could apply the greedy idea efficiently for short reads. The quality of an assembly algorithm is given according to the accuracy and contiguity of the result, and since there is a trade-off between assembly quality measures, it is not a trivial task to compare assembly algorithms.

In addition, the result of any assembly algorithm depends on the dataset used, and each algorithm may perform better in special circumstances. For a new dataset, we cannot choose a better algorithm by just comparing previous results on another dataset. There is also another problem, that of the use of parameters, which have to be predefined by the user, in some algorithms (like parameter k in de Bruijn graph-based methods), which makes the comparison more difficult. The final assembly result is definitely dependent on the parameter chosen for the assembly task. There are some metrics available for comparing assembly algorithms, but the availability of a good metric that is not dependent on the reference genome is still missing from the literature.

References

1. Myers, E. W., et al. (2000). A whole-genome assembly of Drosophila. *Science, 287*(5461), 2196–2204.
2. Li, R., et al. (2010). De novo assembly of human genomes with massively parallel short read sequencing. *Genome Research, 20*(2), 265–272.
3. Simpson, J. T., et al. (2009). ABySS: A parallel assembler for short read sequence data. *Genome Research, 19*(6), 1117–1123.
4. Almeida, N. F., et al. (2009). A draft genome sequence of Pseudomonas syringae pv. tomato T1 reveals a type III effector repertoire significantly divergent from that of Pseudomonas syringae pv. tomato DC3000. *Molecular Plant-Microbe Interactions, 22*(1), 52–62.
5. Green, S., et al. (2010). Comparative genome analysis provides insights into the evolution and adaptation of Pseudomonas syringae pv. aesculi on Aesculus hippocastanum. *PLoS One, 5*(4), e10224.
6. Rees, D., Husselmann, L., & Celton. J. (2009). De novo genome sequencing of the apple scab (Venturia inaequalis) genome, using Illumina sequencing technology. in PAG-XVII Plant and Animal Genomes XVII Conference. Available online at: http://www.intl-pag.org/17/abstracts/P01_PAGXVII_013.html.
7. Bondy, J., & Murty, U. (2008). *Graph Theory (Graduate Texts in Mathematics vol 244)*. New York: Springer.

8. Smith, T., & Waterman, M. (1981). ªIdentification of common molecular subsequences °. *J. Molecular Biology, 147*, 195–197.
9. Scheibye-Alsing, K., et al. (2009). Sequence assembly. *Computational Biology and Chemistry, 33*(2), 121–136.
10. Pevzner, P. A. (1989). 1-Tuple DNA sequencing: computer analysis. *Journal of Biomolecular Structure & Dynamics, 7*(1), 63–73.
11. Tsur, D. (2010). Sequencing by hybridization in few rounds. *Journal of Computer and System Sciences, 76*(8), 751–758.
12. Dramanac, R., et al. (1989). Sequencing of megabase plus DNA by hybridization: theory of the method. *Genomics, 4*(2), 114–128.
13. Lysov Iu, P., et al. (1988). Determination of the nucleotide sequence of DNA using hybridization with oligonucleotides. A new method. *Doklady Akademii Nauk, 303*(6), 1508–1511.
14. Medvedev, P., et al., Computability of models for sequence assembly. Algorithms in Bioinformatics, 2007: pp. 289–301.
15. Pevzner, P. A., Tang, H., & Waterman, M. S. (2001). An Eulerian path approach to DNA fragment assembly. *Proceedings of the National Academy of Sciences, 98*(17), 9748–9753.
16. Narzisi, G., & Mishra, B. (2011). Comparing de novo genome assembly: The long and short of it. *PLoS One, 6*(4), e19175.
17. Schwartz, D. C., & Waterman, M. S. (2010). New generations: Sequencing machines and their computational challenges. *Journal of Computer Science and Technology, 25*(1), 3–9.
18. Ariyaratne, P. N., & Sung, W. K. (2011). PE-Assembler: De novo assembler using short paired-end reads. *Bioinformatics, 27*(2), 167–174.
19. Warren, R. L., et al. (2007). Assembling millions of short DNA sequences using SSAKE. *Bioinformatics, 23*(4), 500–501.
20. Dohm, J. C., et al. (2007). SHARCGS, a fast and highly accurate short-read assembly algorithm for de novo genomic sequencing. *Genome Research, 17*(11), 1697–1706.
21. Jeck, W. R., et al. (2007). Extending assembly of short DNA sequences to handle error. *Bioinformatics, 23*(21), 2942–2944.
22. Ewing, B., & Green, P. (1998). Base-calling of automated sequencer traces usingPhred. II. error probabilities. *Genome Research, 8*(3), 186–194.
23. Batzoglou, S., et al. (2002). ARACHNE: A whole-genome shotgun assembler. *Genome Research, 12*(1), 177–189.
24. Miller, J. R., et al. (2008). Aggressive assembly of pyrosequencing reads with mates. *Bioinformatics, 24*(24), 2818–2824.
25. Huang, X., & Madan, A. (1999). CAP3: A DNA sequence assembly program. *Genome Research, 9*(9), 868–877.
26. Margulies, M., et al. (2005). Genome sequencing in microfabricated high-density picolitre reactors. *Nature, 437*(7057), 376–380.
27. Hernandez, D., et al. (2008). De novo bacterial genome sequencing: millions of very short reads assembled on a desktop computer. *Genome Research, 18*(5), 802–809.
28. Hossain, M.S., Azimi, N., Skiena, S. (2009). Crystallizing short-read assemblies around seeds. *BMC Bioinformatics* 10(Suppl 1), S16.
29. Miller, J. R., Koren, S., & Sutton, G. (2010). Assembly algorithms for next-generation sequencing data. *Genomics, 95*(6), 315.
30. Zerbino, D. R., & Birney, E. (2008). Velvet: Algorithms for de novo short read assembly using de Bruijn graphs. *Genome Research, 18*(5), 821–829.
31. Chaisson, M. J., Brinza, D., & Pevzner, P. A. (2009). De novo fragment assembly with short mate-paired reads: Does the read length matter? *Genome Research, 19*(2), 336–346.
32. Butler, J., et al. (2008). ALLPATHS: De novo assembly of whole-genome shotgun microreads. *Genome Research, 18*(5), 810–820.
33. Peng, Y., et al. IDBA–a practical iterative de Bruijn graph de novo assembler. in Research in Computational Molecular Biology. 2010. Springer.

34. MacCallum, I., et al. (2009). ALLPATHS 2: Small genomes assembled accurately and with high continuity from short paired reads. *Genome Biology, 10,* R103.
35. Chaisson, M. J., & Pevzner, P. A. (2008). Short read fragment assembly of bacterial genomes. *Genome Research, 18*(2), 324–330.
36. Narzisi, G., & Mishra, B. (2011). Scoring-and-unfolding trimmed tree assembler: concepts, constructs and comparisons. *Bioinformatics, 27*(2), 153–160.
37. Sommer, D. D., et al. (2007). Minimus: A fast, lightweight genome assembler. *BMC Bioinformatics, 8*(1), 64.
38. Huang, X., et al. (2003). PCAP: A whole-genome assembly program. *Genome Research, 13*(9), 2164–2170.
39. Sutton, G. G., et al. (1995). TIGR Assembler: A new tool for assembling large shotgun sequencing projects. *Genome Science and Technology, 1*(1), 9–19.
40. Schmidt, B., et al. (2009). A fast hybrid short read fragment assembly algorithm. *Bioinformatics, 25*(17), 2279–2280.
41. Brockman, W., et al. (2008). Quality scores and SNP detection in sequencing-by-synthesis systems. *Genome Research, 18*(5), 763–770.
42. Pareek, C. S., Smoczynski, R., & Tretyn, A. (2011). Sequencing technologies and genome sequencing. *Journal of Applied Genetics, 52*(4), 413–435.
43. Berglund, E. C., Kiialainen, A., & Syvänen, A. C. (2011). Next-generation sequencing technologies and applications for human genetic history and forensics. *Investigative Genetics, 2*(1), 1–15.
44. Shendure, J., & Ji, H. (2008). Next-generation DNA sequencing. *Nature Biotechnology, 26*(10), 1135–1145.
45. Kircher, M., & Kelso, J. (2010). High-throughput DNA sequencing–concepts and limitations. *BioEssays, 32*(6), 524–536.
46. Novais, R., & Thorstenson, Y. (2011). The evolution of Pyrosequencing® for microbiology: From genes to genomes. *Journal of Microbiological Methods, 86*(1), 1–7.
47. Metzker, M. L. (2009). Sequencing technologies—the next generation. *Nature Reviews Genetics, 11*(1), 31–46.
48. Novák, P., Neumann, P., & Macas, J. (2010). Graph-based clustering and characterization of repetitive sequences in next-generation sequencing data. *BMC Bioinformatics, 11*(1), 378.
49. Shendure, J., et al. (2004). Advanced sequencing technologies: Methods and goals. *Nature Reviews Genetics, 5*(5), 335–344.
50. Dong, H., & Wang, S. (2012). Exploring the cancer genome in the era of next-generation sequencing. *Frontiers of Medicine, 6*(1), 48–55.
51. Mardis, E. R. (2008). Next-generation DNA sequencing methods. *Annual Review of Genomics and Human Genetics, 9,* 387–402.
52. Wash, S., & Image, C. (2008). DNA sequencing: generation next–next. *Nature Methods, 5*(3), 267.
53. Smit, A., R. Hubley, and P. Green, RepeatMasker Open-3.0. 1996–2004. Institute for Systems Biology, 2004.
54. Liu, L., et al., Comparison of Next-Generation Sequencing Systems. Journal of Biomedicine and Biotechnology, 2012. 2012.
55. McNally, B., et al. (2010). Optical recognition of converted DNA nucleotides for single-molecule DNA sequencing using nanopore arrays. *Nano Letters, 10*(6), 2237–2244.
56. Hui, P., Next generation sequencing: chemistry, technology and applications. [Without Title], 2012: pp. 1–18.
57. Eid, J., et al. (2009). Real-time DNA sequencing from single polymerase molecules. *Science, 323*(5910), 133–138.
58. Clarke, J., et al. (2009). Continuous base identification for single-molecule nanopore DNA sequencing. *Nature Nanotechnology, 4*(4), 265–270.
59. Tyagi, S., et al., Molecular beacons: hybridization probes for detection of nucleic acids in homogeneous solutions. Nonradioactive Analysis of Biomolecules, 2nd ed. C. Kessler, ed. Springer-Verlag, Berlin, 2000: pp. 606–616.

60. Morozova, O., & Marra, M. A. (2008). Applications of next-generation sequencing technologies in functional genomics. *Genomics, 92*(5), 255–264.
61. Tammi, M. T., et al. (2003). Correcting errors in shotgun sequences. *Nucleic Acids Research, 31*(15), 4663–4672.
62. Paulsen, I. T., et al. (2002). The Brucella suis genome reveals fundamental similarities between animal and plant pathogens and symbionts. *Proceedings of the National Academy of Sciences, 99*(20), 13148–13153.
63. Wu, M., et al. (2004). Phylogenomics of the reproductive parasite Wolbachia pipientis wMel: A streamlined genome overrun by mobile genetic elements. *PLoS Biology, 2*(3), e69.
64. Gill, S. R., et al. (2005). Insights on evolution of virulence and resistance from the complete genome analysis of an early methicillin-resistant *Staphylococcus aureus* strain and a biofilm-producing methicillin-resistant *Staphylococcus epidermidis* strain. *Journal of Bacteriology, 187*(7), 2426–2438.
65. Baba, T., et al. (2002). Genome and virulence determinants of high virulence community-acquired MRSA. *The Lancet, 359*(9320), 1819–1827.
66. Eppinger, M., et al. (2006). Who ate whom? Adaptive Helicobacter genomic changes that accompanied a host jump from early humans to large felines. *PLoS Genetics, 2*(7), e120.
67. Blattner, F. R., et al. (1997). The complete genome sequence of *Escherichia coli* K-12. *Science, 277*(5331), 1453–1462.

Index

A
Assembly errors, 50, 51, 78

C
Chargaff's rule, 2
ChIP-Seq, 35
Coverage, 7, 12, 43–46, 51, 55, 57, 58, 61, 68–70, 72, 75, 78

D
Data preprocessing, 51
De Bruijn graph, 57–59, 71, 75, 79
De novo assembly, 41, 42, 48, 51, 70
De novo genome sequence assembly, 42
De novo Sequencing, 46

E
Emulsion PCR, 13, 17, 21
Error correction, 48, 50, 69, 71, 74

F
Finishing, 46, 49, 56

G
Gap filling, 7, 56, 74
Greedy algorithms, 52, 60, 61, 79

H
Heliscope, 23

I
Illumina/ Solexa genome analyzer, 11
Ion semiconductor, 12, 19

K
k-mer, 44, 47, 58, 59, 61, 63, 69–71, 74

M
Map based sequencing, 6
Mate paired, 7
Maxam-Gilbert, 3, 4
Metagenomics, 36

N
N50, 50, 51, 70, 75, 78
Nanopore sequencing, 12, 25

O
Overlap graph, 57–59, 66, 68, 79
Overlap layout consensus (OLC) algorithms, 59, 66

P
Paired end, 7, 21, 33
Phred score, 32
Polonator, 12, 21
Pyrosequencing, 12, 17, 37, 68

Q
Quality value, 44, 51

A. Masoudi-Nejad et al., *Next Generation Sequencing and Sequence Assembly*,
SpringerBriefs in Systems Biology, DOI: 10.1007/978-1-4614-7726-6,
© The Author(s) 2013